the GIS 20 essential skills

GINA CLEMMER

second edition

Esri Press
REDLANDS | CALIFORNIA

Esri Press, 380 New York Street, Redlands, California 92373-8100

17 16 15 14 13   1 2 3 4 5 6 7 8 9 10

Library of Congress Cataloging-in-Publication Data
Clemmer, Gina, 1974–
    The GIS 20 : essential skills / Gina Clemmer. —Second edition.
        pages cm
    Includes index.
    ISBN 978-1-58948-322-4 (pbk. : alk. paper) 1. Geographic information systems. 2. ArcGIS. I. Title.
    G70.212.C58 20113
    910.285   dc23   2013009216

Ask for Esri Press titles at your local bookstore or order by calling 800-447-9778, or shop online at esri.com/esripress. Outside the United States, contact your local Esri distributor or shop online at eurospanbookstore.com/esri.

Esri Press titles are distributed to the trade by the following:

In North America:
Ingram Publisher Services
Toll-free telephone: 800-648-3104
Toll-free fax: 800-838-1149
E-mail: customerservice@ingrampublisherservices.com

In the United Kingdom, Europe, Middle East and Africa, Asia, and Australia:
Eurospan Group
3 Henrietta Street
London WC2E 8LU
United Kingdom
Telephone: 44(0) 1767 604972
Fax: 44(0) 1767 601640
E-mail: eurospan@turpin-distribution.com

CONTENTS

A NOTE TO THE READER

Over the past decade, it has been my passion to provide an applied approach to teaching GIS to busy professionals. My pedagogy is different from others. I believe the fundamentals of GIS (and ArcMap specifically) can be taught

- quickly. I do not believe that GIS is the deeply complicated discipline many make it out to be.

- using a project-based paradigm by completing a concrete task as the goal, versus a layered or building-block approach. In other words, I don't think you need to know the inner workings of the GIS to create common types of maps.

- by learning what is most frequently used by most people (and therefore most important to understand), and skipping the rest. I do not believe you must know every aspect of ArcMap in order to successfully complete common GIS tasks.

- using everyday language versus technical jargon.

This book is an extension of my passion to help professionals new to GIS quickly learn the essential fundamentals of ArcGIS. The purpose of the book is to provide a focused approach to learning GIS by offering clear, easy-to-follow exercises for 20 of the most commonly used GIS skills in the industry today.

I have spent the past decade training thousands of new GIS users. I created a workshop for busy professionals called Mapping Your Community: An Introduction to GIS and Community Analysis. It has been taken by more than 20,000 working professionals. At the end of each class, students are asked to provide feedback about the course and its content. These student evaluations helped shape my class framework and contributed mightily to the content of this book.

In addition to the feedback my students offer, in October 2008 I conducted a survey of 500 GIS professionals to determine and quantify the top 20 GIS skills used by working professionals. This book is a direct result of that research. The chapters that follow reflect the survey's results, as well as my empirical knowledge of the GIS industry.

Gina Clemmer
New Urban Research Inc.
June 2013

INTRODUCTION

You can read this book in sequence or skip to the chapters that target the skills you want to acquire. How to obtain data to use in the exercises is discussed in the beginning of each chapter.

To get started using this book, you need ArcGIS for Desktop software and data. Instructions on how to get both are given below.

ArcGIS 10.1 for Desktop

You must have an installed copy of ArcGIS for Desktop (any license) to complete these exercises. Many other companies make GIS applications, but given the popularity and functionality of ArcGIS software, we chose this software to illustrate the GIS 20.

Setting up an Esri global account

You will not get far without an Esri global account. The account is free and is required to download the 180-day-use version of ArcGIS 10.1 for Desktop (this book provides an authorization number on the inside back cover to access the 180-day-use version of the software).

If you already have a global account, skip this step. If you are not going to download the software, skip this step.

To create a free global account, do the following:

1. Open an Internet browser and navigate to http://www.esri.com.

2. In the search field (upper right corner) type **Esri Global Account**.

3. Click the first link and select the option "Create An Account." (Links change frequently; however, here is a link for that page http://www.webaccounts. esri.com

Downloading ArcGIS 10.1 for Desktop software (180-day use)

A free, fully functioning 180-day-use version of ArcGIS 10.1 for Desktop software, Advanced license level, can be downloaded at http://www.esri.com/ thegis20forArcGIS10-1. You will find an authorization number printed on the inside back cover of this book. You will use this number when you are ready to install the software.

If you already have a licensed copy of ArcGIS 10.1 for Desktop installed on your computer (or have access to the software through a network), do not install the 180-day-use software. Use your licensed software to do the exercises in this book. If you have an older version of ArcGIS installed, you must uninstall it before you can install the 180-day-use software.

The 180 days begin when you install and register the software. It is a good idea not to register it until you are ready to use it. It will become inactive 180 days after registration, regardless of whether you have ever used it.

To download and install a free 180-day-use version of ArcGIS 10.1 for Desktop, do the following:

1. Check the system requirements for ArcGIS to make sure your computer has the hardware and software required for the trial: **http://www.esri.com/arcgis101sysreq**.

2. Uninstall any previous versions of ArcGIS for Desktop on your computer.

3. Go to **http://www.esri.com/thegis20forArcGIS10-1**, and then follow the instructions for obtaining the software.

Note: When prompted, enter your 12-character authorization number (EVAxxxxxxxxx) printed on the inside back cover of this book. The DVD included with this book does not contain the software, only data files used to complete the chapters. Assistance, FAQs, and support for your 180-day-use software are available on the online resources page at **http://www.esri.com/trialhelp**.

Licensing

Three licensing levels are offered for ArcGIS 10.1 for Desktop. The language is new for this version of the software:

- ArcGIS 10.1 for Desktop Basic was previously known as ArcView level licensing.

- ArcGIS 10.1 for Desktop Standard was previously known as ArcEditor.

- ArcGIS 10.1 for Desktop Advanced was previously known as ArcInfo.

- ArcGIS is now ArcGIS <version> for Desktop to distinguish it as the desktop GIS product.

Language

The GIS 20 explains ideas and steps in nontechnical, everyday language using terms that the average person can understand, not GIS jargon. Below are some common points of confusion.

What is the difference between ArcGIS, ArcMap, and ArcView?

ArcGIS, ArcMap, and ArcView are often used interchangeably. This book uses the terminology "ArcMap" during the exercises, because it is short and easy to reference. ArcMap is actually the software's browser; it is not the whole software. (Just as Internet Explorer is an Internet browser, not the Internet itself.)

"ArcGIS 10.1 for Desktop" refers to the entire mapping software suite available for use on a PC. Years ago this software was called ArcView, but in 1999 Esri made many enhancements and changed the name to ArcGIS; it has been changed again recently to ArcGIS for Desktop to distinguish it from ArcGIS for other platforms, such as ArcGIS for Server and ArcGIS for Mobile.

The exercises in this book can be done using the most basic version of the software, which is ArcGIS 10.1 for Desktop Basic. Other licenses give access to other extensions used for specialized tasks, but these are not used in this book. Keep in mind that the software that you can download with the access code provided in this book is a higher-level license, so you will have the opportunity to see and use some of the extensions and advanced functions.

What is the difference between shapefiles, layers, and .shp files?

In everyday language, shapefiles are called many different things: layer files, boundary files, geography files, and of course shapefiles. This book uses the terms "shapefile" and "layer" interchangeably. This might be confusing because there is actually a layer file type (.lyr).

Layer (.lyr) files are covered in bonus exercise 1 and mentioned in chapter 20. When they are used, it will be clear that we are discussing a layer file type, not a shapefile. Assume every file is a shapefile unless it is specifically identified as a different file type.

Also, sometimes the full name of the shapefile is used, such as "states.shp," but then later in the text the name is shortened to "states," or it is referred to generally as the "states layer" or "states shapefile." This all refers to the same file, so try not to be confused.

Data

All data needed to perform the exercises in this book is included on the accompanying DVD. Most chapters do not require you to use the DVD; however, a few, such as chapter 8 ("Address mapping"), do. Files are provided if you would like an easy, one-source solution to finding files to complete these exercises. In real life, it is not often someone hands you a file with everything you need. With this in mind, the chapters are written to teach you to download and customize files for your purposes, as you might do in your real working life.

At the beginning of each chapter, the hard-drive location of the files needed for that chapter is given, for example C:\EsriPress\GIS20\01. All files identified as being in the C:\EsriPress\GIS20\ folder are from the accompanying DVD. File locations are not given every time the files are mentioned.

DVD installation

To install the DVD that comes with this book, do the following:

1. Put the data DVD into your computer's DVD drive. A splash screen will appear.

2. Click the Install Resources link. This launches the install wizard. Click Next.

3. Accept the license agreement terms, and then click Next.

4. Accept the default installation folder (C:\EsriPress) or click Browse and navigate to the drive or folder location where you want to install the data. All references to the location of DVD files in this book reference the C:\EsriPress folder. Click Next.

5. The installation will take a few moments, and a message will notify you when the installation is complete. When the installation is complete, click Finish.

6. The exercise data is installed on your computer in a folder called EsriPress.

Tricks of the trade

This section covers those things that someone who has been working with ArcMap for a long time *just knows*. Every profession has them, tricks of the trade, things that will make your chances of success better. The "tricks" below are things you really must know in order to be happy and successful with this software.

Finding files

Steps for finding files are not repeated in each chapter of this book. Who wants to repeat themselves constantly? Not I.

So here is what you need to know. Read the "connect to folder" discussion in chapter 1 carefully. You must first "connect to folder" in order to locate and connect to your files. Novices who have not read the section you are reading now spend a tremendous amount of time and frustration trying to find their files. Additionally, various file types are visible in different places, adding to the confusion. For example, shapefiles (.shp) cannot be viewed from the Open menu, and ArcMap documents (.mxd) cannot be viewed when using the Add Data tool. It is worth it to have a look at bonus 1, "Understanding common file types," to get the gist of how various file types are saved and then reopened.

If you become very comfortable navigating ArcMap and saving and opening your files, not just *kind of* comfortable, or *sort of* comfortable, but *deeply* comfortable doing this, you are guaranteed to learn this software more quickly and with less frustration.

Right-click > Properties

If one tip can make your life easier, it is to right-click the layer name in the table of contents, and then click Properties. Shorthand such as "right-click > Properties" is often used to describe this function. This will give you access to all available options for that layer.

This "right-click > Properties" trick also works with almost all items in ArcGIS 10.1 for Desktop. "Right-click > Properties" is the answer to almost every "How do I...?" question in ArcMap. If you are ever in doubt as to what to do, click the item to make it active, and then right-click > Properties. This will generally get you in the right neighborhood.

Toolbars

This software comes with 46 toolbars encompassing hundreds of tools. Toolbars can be accessed in ArcMap by going to the Customize menu and clicking Toolbars. There are three toolbars you will always use:

Standard

Tools

Draw

It is a good idea to make sure these are enabled before starting your work. Once they are enabled, they stay enabled until you turn them off.

You are also able to create customized toolbars. If there are certain sets of tools you use repeatedly, then this is a great option.

Help menu

Since this book does not cover every aspect of ArcGIS 10.1 for Desktop, learning to use the Help menu will be, well, helpful. Two Help menus are built into the ArcMap interface. ArcGIS for Desktop Help comes with the software and is a static body of work, meaning that it is not updated as time goes on. The second Help option is the ArcGIS Resources website (formerly called the ArcGIS Resource Center). The great thing about the ArcGIS Resources site is that it is frequently updated and loaded with new information. Both are accessible from the ArcMap window under Help (upper, right).

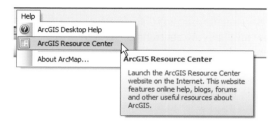

Internet browsers

The official recommendation for ArcGIS 10.1 for Desktop compatible Internet browsers is Internet Explorer 8.0.

ArcGIS 10.1 for Desktop: What's new?

ArcGIS 10.1 for Desktop includes many key enhancements to geoprocessing, labeling, and sharing maps. Things work more smoothly and it is easier to quickly create and share maps.

Below are the top 10 updates introduced in ArcGIS 10.1 as presented at the Esri International User Conference:

10. Searching: Searching is improved throughout the entire software. A good example is the ability to search for coordinate systems based on an opened map. The search is conducted for what is displayed in the map frame.

9. Editor tracking: This function allows you to make a log of who edited data and when. Each time an edit is made, information about that transaction is logged.

8. Geotagged Photos to Points: This new tool enables you to snap photos with a geoenabled device (such as certain digital cameras and smartphones), quickly upload the points to ArcMap, and then click the points to see attached photos. The Geotagged Photo to Points tool can be found in ArcToolbox > Data Management Tools > Photo.

7. GPX to Features: Many GPS units produce outputs in the GPX file format. The GPX to Features tool makes it very easy to upload points in this file format. This tool can be found in ArcToolbox > Conversion Tools > From GPS.

6. Better KML support: ArcMap now displays symbols, labels, and even pictures from imported KML files.

5. Multiscale maps (Advanced license): The Generalization toolset now includes two new tools to create multiscale maps, Collapse Road Detail and Delineate Built-Up Areas. These tools can be found in ArcToolbox > Cartography Tools > Generalization.

4. Geodatabase Administration: A new tool is available to help GIS administrators control permissions and management of geodatabases.

3. Key Indexing (Labels): Part of the new Maplex label functionality, Key Indexing allows users to use indexes to indicate what labels should be in areas where the geography is too dense to place the labels.

2. Dynamic legends: Legends in ArcMap now change based on the scale of the map. As you zoom in or out, the legend changes to accommodate what is displayed in the map frame.

1. Share As command: This command allows you to publish maps easily to your server, other servers, or to your desktop. The Share As command can be found on the File menu in ArcMap.

CHAPTER 1

Downloading shapefiles and using essential ArcMap tools

Shapefiles are the building blocks of many maps and a natural place to begin. Shapefiles are layers that can be stacked on top of each other, like a layered cake, to create one composite map image. Shapefiles contain both the map and the underlying data for the map. In your GIS life, and in this book, you will need various types of shapefiles. You will also need an understanding of essential tools and key ArcMap features such as the table of contents.

In this exercise, you will download shapefiles from the census and customize them to suit your purposes. You will also learn essential tools and key ArcMap features.

Files and tools

Files needed: Files for this exercise will be downloaded from the Internet. Or, if you prefer, you can install this book's DVD and access chapter files at C:\EsriPress\ GIS20\01. Don't know how to install the DVD? See "DVD installation" on page xii.

Tools needed: ArcGIS 10.1 for Desktop software, an Internet connection and browser, and a file-unzipping program.

Downloading shapefiles

You should look at three things when downloading shapefiles: 1. how current the shapefile geography is, 2. how much it costs to obtain the shapefile, and 3. how accurate the shapefile is. The US Census Bureau is an excellent resource for shapefiles. The files are fairly current, free, and pretty accurate.

The US Census Bureau is the national custodian of geographic definitions of US borders. It is the largest distributor of free shapefiles. The Bureau previously maintained geographic boundaries in TIGER/Line file format but several years ago switched to shapefile format. Shapefiles are updated at least once a year and are as current as the previous year.

In the following chapters we will need shapefiles to create maps. Two useful shapefiles are counties and cities. Let's download these two now. But first, we need a place to save files.

1 Set up a Save folder

1. Using Windows Explorer (outside of ArcMap), create a folder on your C drive, such as C:\GIS20, where you will save files for this exercise and others. Going forward, this folder will be referred to as your "save folder."

2 Select files from the US Census Bureau website

1. Navigate to the US Census Bureau website, **http://www.census.gov**.

2. On the Census Bureau site, at the top of the page, click Geography, and then click the TIGER link.

3. Click the TIGER/Line Shapefiles link.

TIGER Products	Which product should I use?						
	Product	Best For...	File Format	Type of Data	Level of Detail	Descriptive Attributes	Vintages Available

• Partnership Shapefiles
• Relationship Files
• Comparability Files
  • Places
  • County Subdivisions
• Gazetteer Files
	TIGER/Line Shapefiles	Most mapping projects--this is our *most comprehensive dataset*. Designed for use with GIS (geographic information systems).	Shapefiles (.shp) and database files (.dbf)	Boundaries, roads, address information, water features, and more	Full detail (not generalized)	Extensive	2006 - 2012, CD 113

• Block Assignment Files • Name Lookup Tables • Tallies • LandView	TIGER/Line Shapefiles & Geodatabases with Demographic Data	Demographic analysis from selected attributes from the 2010 Census, 2006-2010 ACS 5- year estimates, and 2007-2011 ACS 5-year estimates for selected geographies. Designed for use with GIS.	Shapefiles (.shp) and Geodatabases	Boundaries, Population Counts, Housing Unit Counts, 2010 Census Demographic Profile 1 attributes, 2006-2010 ACS 5-year estimates data profiles, 2007-2011 ACS 5-year estimates data profiles.	Full detail (not generalized)	Limited	2010, 2006-2010 ACS, 2007-2011 ACS
	Cartographic Boundary Files	Small scale (limited detail) mapping projects clipped to shoreline. Designed for thematic mapping using GIS.	Shapefiles (.shp)	Selected boundaries	Less detail (generalized)	Limited	2010, 2000, 1990
	KML Prototype Files	Viewing data or creating maps using Google Earth, Google Maps, or other platforms that use KML.	KML (.kml)	Selected boundaries	Less detail (generalized)	Limited	2010
	TIGERweb	Viewing maps online or streaming to your mapping application.	interactive viewer and WMS	Boundaries, roads, address information, water features, and more	Detailed and generalized	Extensive	2010

4. Click the 2011 TIGER/Line tab. These files are updated annually. Download the most current year.

5. Next, click the Download link, and then click the Web interface link.

3 Select a layer type

First you must select what type of geographic file to download.

1. From the "Select a layer type" list, click Counties (and equivalent), and then click the Submit button. →

2. On the next page, click the Download National File button. The County shapefile will begin to download. This file includes all counties in the United States; it is a large file and may take a while to download.

3. Depending on which Internet browser you are using, you may be asked whether you want to Open, Save, or Cancel the file. Click Save to save the

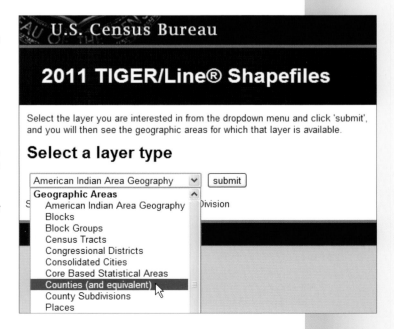

U.S. Census Bureau

2011 TIGER/Line® Shapefiles

Select the layer you are interested in from the dropdown menu and click 'submit', and you will then see the geographic areas for which that layer is available.

Select a layer type

American Indian Area Geography ▼ [submit]

Geographic Areas
American Indian Area Geography
Blocks
Block Groups
Census Tracts
Congressional Districts
Consolidated Cities
Core Based Statistical Areas
Counties (and equivalent)
County Subdivisions
Places

zipped file to your C drive. If you are not asked this, select whichever options are necessary to save the zipped file to your save folder.

4. Once the file downloads, a blank screen will appear. This looks like an error, but it is not. Click the back arrow twice to return to the "Select a layer type" list.

5. Click Places, and then click Submit.

6. Next, select the state in which you live (this state will be used in many subsequent exercises), and then click Download. In this example, Alabama will be selected (just because it is the first state on the list). Save the file to your save folder.

7. Unzip the files. You can use any unzipping program you wish. Follow the instructions for that program to unzip the files. You will see several files after you unzip. If you are unsure how to unzip, try right-clicking the zipped file and extracting, or extracting all. You might also try double-clicking.

Adding shapefiles to ArcMap

4 Open ArcMap

1. On the Windows Start menu, click All Programs, ArcGIS, and ArcMap 10.1. The software will take a moment to open.

2. If a window appears asking you to select a blank map template, click "Blank Map" (or you can click the Cancel button). If this window does

US CENSUS GEOGRAPHIES CAN BE CONFUSING

The US Census Bureau website allows you to select shapefiles and tabular data for many different types of geography (tracts, counties, the entire nation, states, etc.). Here is a quick reference to the most widely used geographies:

- Nation: This is for the United States as a whole. If you select this geography and then, for example, population as a data variable, the result will be one number, the population of the entire United States.
- State: Allows you to select one state, multiple states, or all states.
- County: Allows you to select one county, multiple counties, or all counties for the entire United States.
- Place: Represents city boundaries, plus Census Designated Places.
- Census tract: Tracts are the most popular subcounty geography. They are fixed in population between 1,000 and 8,000 people. Census tracts average about 4,000 people, although this varies widely.

WHAT FILES MAKE UP A SHAPEFILE?

After unzipping the Census files, you will have a file with the .shp extension, but you will also have a few other files with different file extensions. That is because a shapefile is actually made up of multiple files, not just the .shp file.

Mandatory files

    .shp = the visual image of the map

    .dbf = database file, where data is warehoused for the shapefile

    .shx = index file that ties the .shp and .dbf files together

Optional files

    .prj = projection file, gives your map its shape, area, direction, and distance (discussed in chapter 3)

    .shp.xml = contains metadata about your shapefile

    .sbn and .sbx = spatial index files

    other files = there can be a few other file types associated with a shapefile

not appear, it is not a problem. You are simply trying to get to a new empty mapping session.

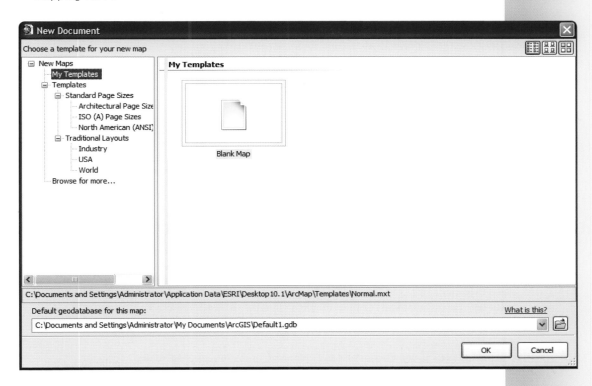

NOTE: Step 4, "Open ArcMap," is not repeated throughout this book.

5 Add shapefiles

1. Add shapefiles to ArcMap by clicking the Add Data button 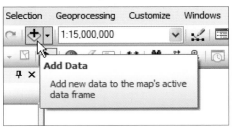 You will use this tool constantly, so it might be a good idea to get to know it (see the "Finding files" discussion on page xii). →

When beginning work on a project, it is helpful to "connect to folder" the first time you access your files. By connecting to folder, you don't have to navigate through a maze of links every time you want to add a shapefile or open a project. A permanent shortcut is created under the Folder Connections section. (If you haven't already done so, create a save folder on your C drive). →

2. Select the Connect to Folder button 🗁 and navigate to the created folder (in this example C:\EsriPress\GIS20\01). Click OK.

3. Hold down the Ctrl key, select the unzipped Census shapefiles tl_2011_us_county.shp and tl_2011_01_place.shp, and click the Add button.

> **NOTE:** Unless you also downloaded Alabama, you will have slightly different file names. Your state's FIPS code will replace the "01" in the file name.

If you do not see shapefiles here (they have a .shp extension), it is likely you have not properly unzipped the files. Go back and try to unzip and add them again.

FIPS CODES

FIPS code means Federal Information Processing Standard code, which provides a unique ID for every parcel of land in the United States. States have two-digit codes, and counties have three. So a state plus a county code is a five-digit unique identifier for every county in the United States.

HOW DO I FIND MY FILES?

When you first open ArcMap and click the Add Data button to add shapefiles or other files to ArcMap, the C drive will not be accessible via the navigation list. This has led to many a frustrated beginner.

The answer here is you must first "connect to folder" in order to access the C drive. There are two ways to connect to folder:

1. Through the Add Data tool and the Connect to Folder tool 🗁.

2. Through ArcCatalog 🗊 and the Connect to Folder tool, or by right-clicking Folder Connections .

Also, you do not actually have to choose a folder; you can just navigate to the desktop and connect there. All connected pathways should be evident under the Folder Connections link via the Add Data tool or through ArcCatalog.

6 Explore essential tools

Because ArcMap provides hundreds of tools, it is essential to identify the most important ones. The tools you will use in nearly every mapping session are featured below. It is a good idea to become very familiar with these tools and what they can do.

Add Data ⊕ ▾

This one you already know because we just used it, but it does not hurt to reiterate here. The fastest way to add geography files, layer files, or data tables is to click the Add Data button on the Standard toolbar.

1. Click the Add Data button and get comfortable with using the navigation list to add files, connect to folders, and generally just find your way around. You will notice if you click the little down arrow next to the button that other options appear. We will get to those other options later, for now we will use Add Data.

Zoom In/Zoom Out, Fixed Zoom In/Fixed Zoom Out

ArcGIS has four tools for zooming in and out of your map: Zoom In, Zoom Out, Fixed Zoom In, and Fixed Zoom Out.

2. Click the Zoom In button ⊕. The tool works better by drawing a square around whatever you want to magnify instead of just clicking the map. For example, activate the tool, then draw a square with it around your state. Notice how it is easier to control the image by first drawing a square. Now try the other zoom tools and see what happens. Do not worry if you mess up your map. The next tool will help fix it.

> ### INCREDIBLY USEFUL TIP
> *You can also easily zoom in to and out of your map by using the scroll wheel or trackball on your mouse, if you have a mouse with this functionality.*

Full Extent ◉

The Full Extent tool will resize your map so it fits onto your screen. It is a great way to center your map.

3. Click the Full Extent button and notice how your map is repositioned.

Pan ✋

The Pan tool looks like a little hand. It allows you to reposition the map as if you were moving it on your screen with your hand.

4. Click the Pan button and move your map around. To recenter it, click the Full Extent button again.

Default pointer ▶

The default pointer doesn't do anything. That's the beauty of it.

> ### INCREDIBLY USEFUL TIP
> *To "deactivate" any of the other tools, click the default pointer. This gets rid of the first tool and activates the default pointer.*

5. Click the default pointer, and then click your map. Notice nothing happens.

Identify

Identify is one of the most useful tools. You can use the Identify tool to click specific geographies and look at the underlying data. Use the Zoom In tool to zoom in closely to a few counties in your state.

6. Click the Identify tool, and then click a county. Notice that a box with county information appears. Try a few more counties until you are comfortable with this tool. Click the default pointer to get rid of the Identify tool.

7 Explore the table of contents

The Table of Contents window, located on the left side of the ArcMap window, is the organizational panel for working with files in ArcMap. Notice there are two shapefiles listed in the table of contents under Layers shown in the illustration. Also notice the five buttons at the top of the table of contents: List by Drawing Order, List by Source, List by Visibility, List by Selection, and Options. Hover over any of these and a description of what that button does displays.

1. Click the first button, List by Drawing Order . The default is the second button, List by Source; however, List by Drawing Order is more useful for most mapping sessions.

2. Use the Zoom In tool to zoom in to your state.

3. Display layers by selecting the box next to each layer to turn it on, or by clearing it to turn it off.

4. Practice moving the county shapefile and the places shapefile. You can move layers up or down by highlighting the layer in the table of contents (clicking it once) and dragging it to the desired position.

> ### INCREDIBLY USEFUL TIP
> *This functionality will only work while in List by Drawing order mode. If you are unable to move these layers up or down, click the first button, List by Drawing Order, and try again.*

Notice how your map changes as you reposition the layers. When the places layer is on top, you are able to see cities in your state. When the county layer is on top, it blocks out the places layer because it is the top layer and has a solid fill color associated with it. Move the place layer into first position.

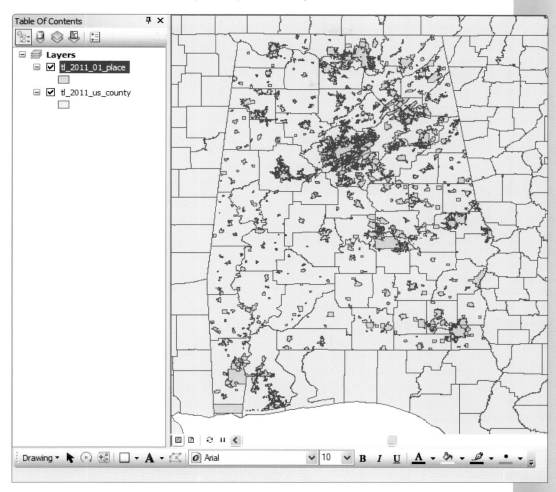

The layer names are what will be used in the legend. You may want to make them more reader friendly by using common names such as "Counties" and "Cities."

5. Click the layer name to activate the text. Type over the existing layer name. This does not rename the layer in the underlying data, just in your map. →

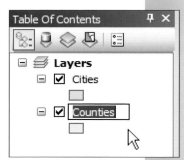

8 Understand shapefiles

Shapefiles contain two things: the map and the underlying data table. So far, we have looked at the map part of the shapefile and used the Identify tool to look at the underlying data. Another way to view the underlying data is to view the whole data table at once.

1. Right-click the county layer name in the table of contents, and then click Open Attribute Table. Use the scroll bars (right, bottom) of the attributes table to better understand what is available in the underlying data table.

You should notice a few thousand counties in the attributes table. There is no demographic data here, only FIPS codes, county name, and some other miscellaneous codes and information the Census Bureau included when they created the shapefile.

2. Right-click any of the column headings, and then click Sort Ascending. Notice that this command sorts the entries alphabetically (if the entry contains text) or from smallest to largest value (if the entry contains numbers). This command is useful when you need to isolate a few entries in a large group.

3. Close the attribute table by clicking x in the upper right corner.

NAME		NAMELSAD	LSAD	CLASSFP	MTFCC
Comerío		Sort Ascending	13	H1	G4020
Aibonito		Sort Descending	13	H1	G4020
Ciales					20
Aguada		Advanced Sorting...			20
Yabucoa		Summarize...			20
Canóvanas		Statistics...			20

Sort Ascending
Sort the values in this field in ascending order (A - Z) (1 - 9).

9 Customize shapefiles

Shapefiles downloaded from the Census, or anywhere, may need to be modified to suit your purposes. A great example of this is when we downloaded the county file. We had no option to download a county file only for our state. The only file given was at the national level, containing all counties for the whole United States.

For the following exercises it would be better to have a shapefile of only counties in your state (like the place shapefile). How do we do that? The Census Bureau does not provide the shapefiles that way. However, it is relatively easy to carve up shapefiles. In the next few steps, let's figure out how to isolate your state's counties and create a new shapefile that contains just those counties.

1. Right-click the places (Cities) shapefile in the table of contents, and then click Open Attribute Table.

2. Find the column titled STATEFP (the third column from the left), and make a note of your state's two-digit FIPS code. For Alabama, this code is 01. Close the attribute table. →

3. Right-click the counties file in the table of contents, and then click Open Attribute Table.

	FID	Shape	STATEFP	PLACEFP	PLACENS
	0	Polygon	01	15640	02404082
	1	Polygon	01	24568	02403575
	2	Polygon	01	44344	02406049
	3	Polygon	01	03724	02405200
	4	Polygon	01	07672	02405286
	5	Polygon	01	15376	02406274
	6	Polygon	01	37096	02405877
	7	Polygon	01	59472	02404510
	8	Polygon	01	28024	02406517
	9	Polygon	01	55440	02407012
	10	Polygon	01	70128	02407329
	11	Polygon	01	77304	02405617

Cities

4. Right-click the STATEFP column heading, and then click Sort Ascending. This will organize the table by state. Scroll down and find the first record in the STATEFP column that corresponds to your state's FIPS code. Alabama is easy. The code for the state is 01, so the counties in this state are at the very top of the attributes table.

5. Once you find the first record for your state, click the very first gray square at the beginning of the line that contains the first record for your state. This will create a blue highlighted record on that line item. Continue to highlight the remaining records for your state by dragging the mouse down the records (you must be on the beginning gray squares in order to drag highlighting). Highlight all records in your state. →

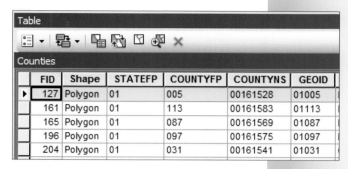

6. It is very important to close the attribute table now. Notice counties in your state are highlighted on the map. The attribute table and map are linked. You may need to zoom in to see your state more closely. →

Now we're going to create a new shapefile for just those counties in your state.

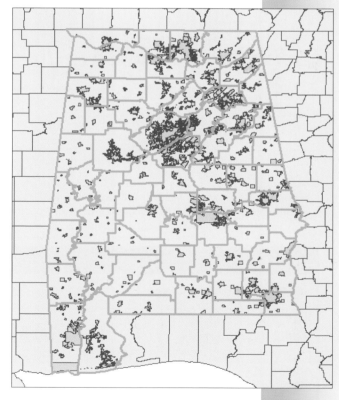

7. Right-click the counties layer name in the table of contents. Click Data, and then click Export Data. Leave all default options here, except click the browse button 📂 and navigate to your save folder (under Folder Connections).

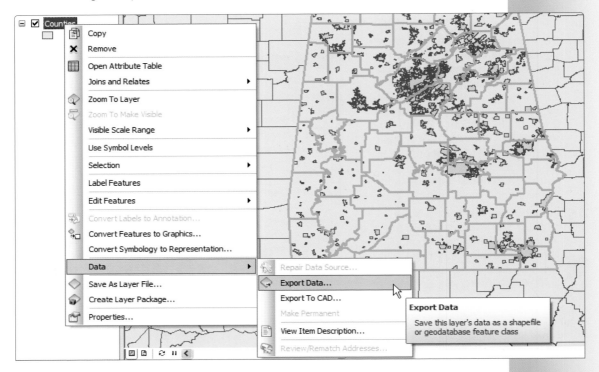

8. Name the file your state's name plus Counties, for example "AlabamaCounties." From the Save as type list, select Shapefile. Click Save, and then OK. When prompted with the box that says "Do you want to add the exported data to the map as a layer?" click Yes. Notice a new layer appears in the table of contents.

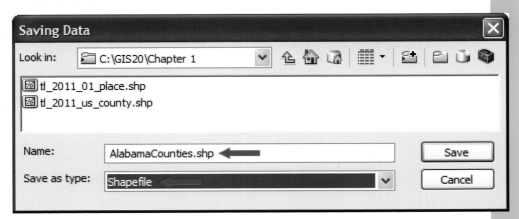

Let's organize things a bit. The US county layer is no longer needed. Let's remove it.

INCREDIBLY USEFUL TIP

Never include spaces in file names.

9. Right-click the US county layer name (this is your original county file) in the table of contents, and then click Remove. Drag the place file into first position in the table of contents. You should see all cities in your state, as well as all counties. →

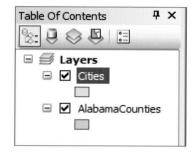

10 Save the project

1. On the File menu, click Save As.

2. Navigate to your save folder.

3. Type a name for this project. Call it **mystate.mxd**.

4. Click Save. Notice now in the upper left corner, your workspace has been given this new name.

5. Click the x in the upper right corner to close ArcMap.

ARCMAP DOCUMENTS (.MXD FILES)

Saving, while technically easy, is more complicated than it appears. When you "save" a project, ArcMap saves the project as an .mxd file or ArcMap document. The .mxd file functions as a pointer file to all the files that make up the project. If you were to send the .mxd file to someone, they would not be able to open it without all the files that made up the project.

CHAPTER 2

KEY CONCEPTS

working with multiple layers
changing map colors
creating labels
creating layouts
creating legends
creating titles
using scale bars
using north arrows

Creating basic maps and layouts

Reference maps are basic, traditional maps like those you see in atlases. Their purpose is to illustrate geographic boundaries of a given area such as cities or counties. These types of maps are the cornerstone of cartography. Layouts contain titles, legends, north arrows, scale bars, and many other graphic features. It is important that you learn to create layouts that help your reader quickly understand your map.

In this exercise, you will create a reference map and layout using fundamental skills such as changing symbology and labeling.

Files and tools

Files needed: You will need mystate.mxd (chapter 1) and alabamacounties.shp (chapter 1). Or, if you prefer, you can install this book's DVD and access chapter files at C:\EsriPress\GIS20\02. Don't know how to install the DVD? See "DVD installation" on page xii.

Tools needed: ArcGIS 10.1 for Desktop.

Creating a reference map

A key point of this exercise is to teach you to effectively color-shade and label multiple layers in a map. In this example, counties and cities are used.

1 Open the project

1. Open ArcMap. On the File menu (upper, left), click Open, and then navigate to your save folder. Select mystate.mxd. (If you are unable to find this file, you can access a similar one at C:\EsriPress\GIS20\02.)

> **RIGHT-CLICK > PROPERTIES**
>
> Right-click > Properties is the answer to almost every "How do I...?" question in ArcMap. Right-clicking a layer name in the table of contents and clicking Properties provides access to Layer Properties, which is like the "control panel" for the shapefile. It contains all options and settings for that shapefile. Right-click > Properties works on most other functions in ArcMap as well.

2 Change layer colors (called "changing the symbology" in ArcMap)

Change the layers' symbology (color, fill, outline) so you can easily see each layer in the map.

For the county layer, follow these steps:

1. In the table of contents, right-click the county file layer name, and then click Properties.

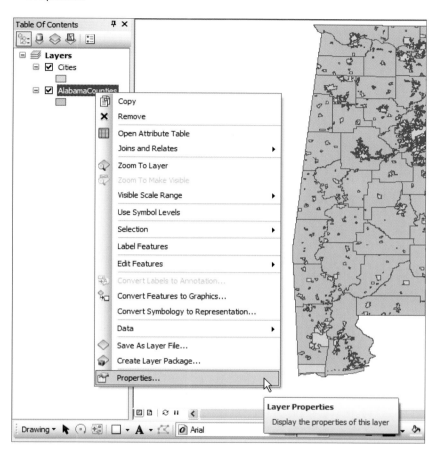

2. Click the Symbology tab. Click the big color swatch under Symbol. ➜

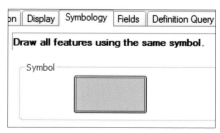

3. Click the hollow color swatch on the left. For the outline color, located slightly below the Fill Color option on the right, select a medium gray. Click OK twice.

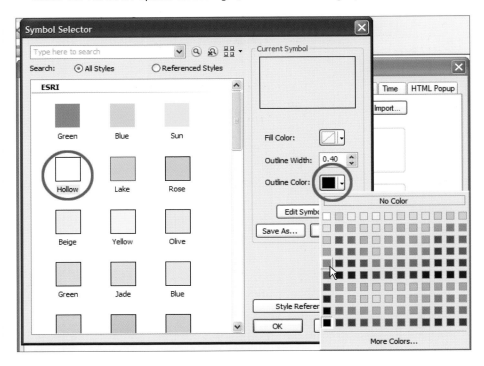

INCREDIBLY USEFUL TIP

"Hollow" makes the layer transparent so you can see the layer underneath it. "Hollow" and "No Color" are one and the same. For many shapefiles, gray is a better choice than black for the outline because it puts less ink on the page, which will make maps easier to read.

For the cities layer, follow these steps:

4. In the table of contents, right-click Cities, and then click Properties. (Assuming you changed the layer name in the last exercise. If you did not, this is the places shapefile.)

5. Click the Symbology tab, and then click the big color swatch under Symbol.

6. Click the Fill Color color swatch on the right side and select a bright green to symbolize cities. A good option here is Light Apple, which is the seventh column over, second row down (hovering over the color swatch displays the color name). Then, for outline color, select No Color. By taking off the outline color, we make the map easier to read. Click OK twice.

3 Make a layer semitransparent

1. Right-click the Cities layer in the table of contents, and then click Properties.

2. Click the Display tab.

3. To the right of Transparent, type **50**. This will make the Cities layer 50 percent transparent, allowing you to see through it. Click OK. →

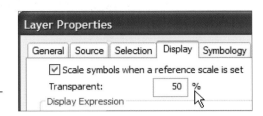

Notice how the layer is lighter. We did not need to make this layer semitransparent; however, it makes the map slightly easier to read and teaches a new skill.

4 Reorganize layers

1. In the table of contents, drag the county layer into first position. If you are unable to do this, make sure List by Drawing Order in the table of contents is selected.

This makes a subtle change. The counties are the top layer, with light gray lines and the semitransparent cities are displayed underneath.

> **INCREDIBLY USEFUL TIP**
>
> *For a reference map, it is difficult to illustrate more than four layers on one map without the map becoming too cluttered.*

5 Turn on labels

Labeling is essential to creating a quality reference map. Let's label cities.

1. In the table of contents, right-click the layer you would like to label (Cities), and then click Properties.

2. Click the Labels tab. Select the "Label features in this layer" box, which is a tiny little check box in the upper left corner. The appropriate field for labeling is already selected by default (Name).

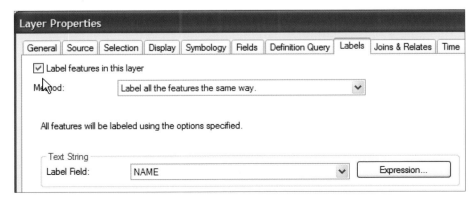

3. Click OK, and have a look at the labels.

What a mess! Let's fix these labels and make the map easier to read.

6 Fix labels

You can do many things to improve messy labels:

- Remove duplicate labels.
- Assign a buffer to labels, which places only some labels.
- Create a halo (a white outline) around labels.

To remove duplicate labels, do the following:

1. In the table of contents, right-click Cities, and then click Properties.

2. Click the Labels tab.

3. Select Placement Properties in the lower left corner. →

4. Click the Placement tab at the top.

5. Select the Remove Duplicate Labels option.

Other Options

Placement Properties...　　Scale Range...

To assign buffers to labels, do the following:

6. Click the Conflict Detection tab at the top.

7. Toward the bottom of the menu there is an option to assign a buffer number. Type **2** in the provided field, and then click OK twice. (This will place only some labels on the map, making it much easier to read.)

To create a halo around labels, do the following:

8. To place a halo around labels, right-click Cities, click Properties, and then click the Labels tab.

9. Click the Symbol button to make changes to label fonts.

Text Symbol

AaBbYyZz

Arial　　　8

B I U　Symbol....

10. Click B to make the label type bold.

11. Click the Edit Symbol button. Edit Symbol...

12. Click the Mask tab and Halo option, and then click OK three times.

The image should look like this one, except with your own state. You may need to resize and reposition it using the zooming and panning tools.

NEW LABELING TOOLS IN ARCGIS 10.1 FOR DESKTOP

The functionality of the Maplex for ArcGIS extension is now part of the core ArcGIS for Desktop product and is called the Maplex Label Engine. This tool, in combination with the existing Label toolbar, provides many more options for labeling. To access this new tool, go to Customize > Toolbars > Labeling, which turns on the Labeling toolbar. On the Labeling toolbar, select Labeling > Use Maplex Label Engine. Then go back to Placement Properties (see step 6.3) and notice many new options. Advanced labeling is covered in bonus exercise 2.

7 Understand the relationship between data and layout views

So far, we have been working in data view. This is where maps are created and where most mapmakers spend much of their time. Alternatively, layout view is where maps are prepared for production by adding a title, legend, north arrow, source, and other layout elements. You might spend 10 percent of your time in layout view (unless you have a really complicated layout), and it is usually the last step before printing.

It is important to understand the relationship between data view and layout view. What you see in layout view is what you have set up in data view. To better understand this relationship, change the size of your map in data view and see how that affects layout view.

1. Use the Fixed Zoom Out tool (or mouse scroll wheel) to make the map much smaller. Click it six or seven times until your map is very small.

2. On the View menu, click Layout View to switch to layout view. Notice that what you see here is a visualization of the data view. These two views are linked.

3. We cannot leave the map like this, so switch back to data view (View > Data View) and use the Full Extent tool ⬤ to resize the map to something more appropriate.

4. Return to layout view (View > Layout View).

8 Create a layout

Depending on whether the geography is horizontal (Nebraska) or vertical (California), change the map's orientation to either portrait or landscape to accommodate the geography shape. To change the orientation from the default, portrait, to landscape, follow these steps:

1. Right-click in the white space anywhere outside of the print area (or what looks like the edge of a piece of paper) in the layout.

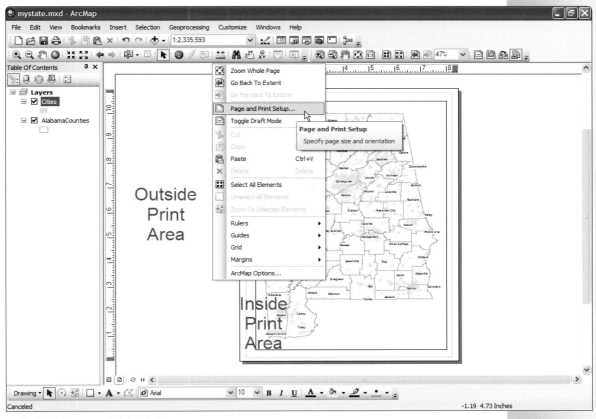

2. Click Page and Print Setup.

3. Select the Landscape option to the right of Orientation.

4. Select the "Scale Map Elements proportionally to changes in Page Size" box (lower right) to center the map in the page layout. Click OK.

5. Resize the image to fit the space by clicking the map once to select it, then dragging the dots in each of the four corners. Do not worry about skewing the map when you drag it—ArcMap keeps the correct proportions of your geography. Click one of the little dots on the corner until a double-headed arrow appears. Drag the image to the desired position, repeat with each corner dot until the map is as large as you would like it.

THE PRINT AREA

In layout view, the print area is defined by the default rectangle that looks like a piece of paper. This rectangle represents an 8 1/2-by-11-inch piece of paper. Stay within the boundaries of this rectangle. Anything outside the boundaries will not print.

6. Optional: Most of the time a landscape page format allows for a larger map image for most geography; however, it is up to you to decide if your state is best suited for landscape or portrait. In this example, Alabama looks better as portrait, so I'm going to switch it back. Feel free to do the same now if you prefer your map to be portrait. You will need to resize it again, by dragging the corner dots again and using the Full Extent tool to resize it.

TOOLS TOOLBAR STILL WORKS

Notice that all the tools on the Tools toolbar are still available and active even in layout view. The Full Extent tool is particularly helpful when working in layout view.

---

9 Insert title

1. In layout view, on the Insert menu, click Title. If the Insert menu is unavailable, it means you are in data view. You must switch to layout view (View > Layout View).

2. Give your map the title **Alabama Counties and Cities, 2011** (except use your state's name).

3. Move the title box above the map by clicking the box with your left mouse button and dragging it to the top of the page. You may need to resize the map frame again to make a little space at the top. To do this, click once on the map to select it. Blue dots will appear around the frame. Move the top part of the frame down a bit to make room for the title.

4. To change the font, click once on the title box to select it. On the Draw toolbar, which might be docked anywhere on your screen, select a larger font size such as 36. Here you can change the font type, size, and style. Change the font size to 36 and make it bold.

This is OK, but the title is a little large for the page. Notice also that it's outside of the print area in between the print margin and the edge. You cannot slightly adjust the font or retype a new title within the Drawing toolbar. For this, we must rely on our old friend right-click > Properties.

5. Right-click the selected title, click Properties (double-clicking the title also works), and click the Change Symbol button.

All font options are contained in the titles properties, making it a much more useful option than using the Drawing toolbar.

6. In the Size box, type **33** instead of 36 to shrink the font size. Click OK twice. You may need to make the title font even smaller.

INCREDIBLY USEFUL TIP

If you need to edit the title, type over the text in the text box that says <dyn type="document" property="title"/>.

FIT TO MARGINS

A quick way to appropriately size a title is to select it, and on the Drawing toolbar, select Drawing > Distribute > Fit to Margins.

MAP FONTS

Of the two types of fonts, serif and sans serif, I believe sans serif is easier to read. Common examples of sans serif fonts include Arial and Helvetica, both of which are smart choices for map fonts. Choose one font for all the text in your map.

10 Insert a legend

1. On the Insert menu, click Legend.

2. Click Next, and then click Next again to proceed through the menu. The default options work fairly well, except for the "predetermining a suitable background color" option.

3. On the Legend Frame wizard, click the drop-down list and select the first option for a border to go around the legend.

4. For the background color, click the menu and select White. The default for a legend is hollow, which makes the legend see-through and awkward to place. A much better choice is a solid fill of white.

5. For a drop shadow, select black (optional). Click Next twice and Finish.

6. Move the legend to one corner, ideally the upper right corner. This is usually a good area for the legend; however, if it obscures too much of your map you will need to find another corner. You will likely need to resize the map frame here and use the Pan tool to reposition the map slightly within the frame.

The shapefile alabamacounties does not have a space between the words. This is great for technical reasons but not great for how it looks on the legend. Also, "Alabama" Counties on the legend is redundant because it already says Alabama in the title.

7. To edit the shapefile name in the legend, click the layer name in the table of contents to make it editable, and then type the name **Counties**. Then click anywhere else to get out of the text editor. Notice how it is changed in the legend. You may need to reposition and size things yet again.

ORGANIZING THE LEGEND

In this map we only have one layer. When working with multiple layers, the legend should be organized in order of scale (from largest to smallest, for example, such as state, county, and city). Another good idea is to put land masses (physical land like state, county, city) at the top of the legend and other context layers, like streets and water, at the bottom.

DYNAMIC LEGENDS, NEW IN ARCGIS 10.1

In ArcGIS 10.1 only those layers that are visible in the map frame populate as the default list under Legend Items, which is a big time-saver. The Legend Properties dialog box has been revamped to provide more helpful options, including legend text wrapping and a Fitting Strategy section under a new Layout tab.

11 Insert a scale bar

Traditional cartography instructs that scale bars must accompany every map; however, scale bars are frequently left off thematic maps (chapter 6) because they serve no real purpose on a thematic map. Scale bars are required on reference maps because distance is an important element.

1. On the Insert menu (top, center), click Scale Bar.

2. Select the first scale bar by clicking it once, and then click OK. If you are using an older version of ArcGIS for Desktop, it will be necessary to modify scale bar properties and change division units from decimal degrees to miles.

The scale bar is dropped right in the middle of the map.

3. Move the scale bar to the lower right corner.

4. The easiest way to change the mileage of the scale bar is to drag the right-side dot of the scale bar box and widen the scale bar. If the map size changes, the scale bar automatically adjusts. It is not a fixed element. Widen the scale bar to show 100 miles.

12 Insert a north arrow

With north arrows, the simpler the better. Avoid overly ornate arrows.

1. On the Insert menu, click North Arrow.

2. The font is difficult to change on ArcMap north arrows. Select a north arrow with a font similar to fonts used in the map, ideally a sans serif one. ESRI North 6 is a good option.

3. Move the north arrow into the legend box and place it in the upper right corner. Make it smaller so it fits in the box.

TRUE NORTH

In ArcGIS 10.1 a true north option has been added in the North Arrow Properties.

13 Insert source using text

Text is a versatile tool and can be used to provide a source. Providing data sources is a must, but you may also consider providing sources for shapefiles as well as the map itself. There are many ways to cite sources. One recommended way is the American Psychological Association (APA) method. Many fields use this method, not just psychology.

1. To insert text, on the Insert menu, click Text.

The text box is dropped in the center of the map, which makes it difficult to see.

2. Click anywhere outside of the text box to deactivate it, and then hover over the text box. A four-pronged arrow will appear. Drag the text box to the lower right corner so it is easier to see and work with.

3. Right-click the text box, and then click Properties (double-clicking also works).

You can have a map source (your agency) as well as a data/geography source (the US Census Bureau).

4. Type the following:

Map Source: Your organization, June 2012.

Shapefile Source: US Census Bureau. (2011, January 1). 2011 TIGER/Line Shapefiles for: Alabama. Retrieved June 28, 2012, from US Census Bureau: **http://www. census.gov/cgi-bin/geo/shapefiles2011/main**.

Notice the font is Arial 10. Arial is a fine sans serif font, but make the font larger.

5. Click Change Symbol and change the font size from 10 to 12. Click OK.

INCREDIBLY USEFUL TIP

Sources, as well as other map text, should generally not be smaller than 10 point font.

---

6. Click the Left Alignment button ▤, to align text left. You must do this step last, otherwise it will not work. Click OK.

Notice the citation is too long and runs off the page. To fix this, do the following:

7. Right-click the text box, and then click Properties.

8. Insert the cursor where you would like to break the line, and then press Enter. Do this four or five times. You can click Apply after each time to see how it looks. Once it looks reasonable, click OK.

14 Save the project

1. On the File menu, click Save As.

2. Navigate to your save folder.

3. Type a name for this project. Call it **refmap.mxd** for reference map. Click Save. Close ArcMap.

MAP FLOW

The philosophy behind map flow is based on the principles of photography. The eye should be drawn to certain anchor points and move across the page. In this example, the eye moves in a Z pattern. Avoid clustering all elements on one side of the map, which creates an unbalanced composition. White space should be fairly equal on all sides of the map.

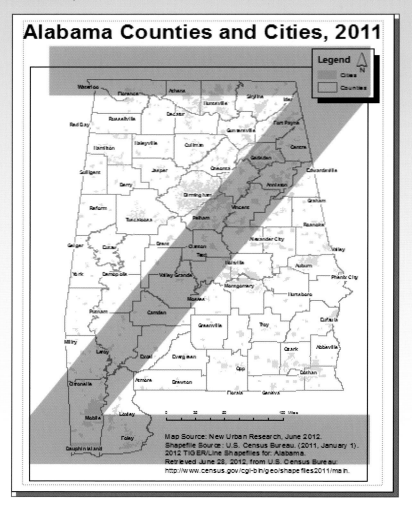

Projecting shapefiles

Projections give shapefiles the correct shape, area, direction, and distance. Defining the projection for shapefiles will ensure that the geography is reflected properly, that distance is recorded accurately, and that all layers are visible. There are easier things to understand than map projections, like quantum physics. This chapter seeks to provide common, and practical, solutions for typical projection issues, not present a treatise on the subject.

In this exercise, you will project shapefiles, using the provided solutions to address common technical issues, and learn clues to help you identify when something is wrong with the projection.

Files and tools

Files needed: You will need alabamacounties.shp (chapter 1). Or, if you prefer, you can install this book's DVD and access chapter files at C:\EsriPress\GIS20\03. Don't know how to install the DVD? See "DVD installation" on page xii.

Tools needed: ArcGIS 10.1 for Desktop.

DISTANCE ACCURACY AND AVOIDING AWKWARD CONVERSATIONS

A good starting point for a discussion on accuracy is to ask how accurate your map needs to be. For basic mapping, especially for demographics and other thematic maps, your map likely does not need to be within a hair's breadth of reality. However, 50 miles off probably won't work either. If distance accuracy is even a little important for your project, consider having a discussion with your team to agree on what projection is appropriate, which could save many awkward discussions later on.

The second thing you should do is make sure that all shapefiles for one analysis are in the same projection so they line up with each other. For example, you may have a layer of bus stops and a layer for streets. You publish this map to the web. Then you look at it closely and realize the stops are not at the correct locations. They are slightly off, or in the worst case, really off. In the meantime, thousands of unsuspecting bus riders looked at your map, and all went to the wrong corners.

Understanding the basics

Coordinate systems can be grouped into **geographic coordinate systems (GCS)** and **projected coordinate systems (PCS)**. The key distinction is whether they use a 3D object (think globe) or a 2D, flat surface (think map) to represent the surface of the earth.

Many coordinate systems are used; however, latitude/longitude is the oldest (developed before Christ) and uses global lines running north-south (latitude) and east-west (longitude) to assign a set of numbers to every location on earth. Latitude/longitude is based on a 3D spherical surface and is the foundation of many coordinate systems.

The problems with projections come when you need to represent a 3D object on a 2D surface. This is where a PCS comes in handy. Universal transverse Mercator (UTM) is a PSC and was developed in the 1940s by the US Army Corp of Engineers. It uses a grid system to assign numbers to grid cells worldwide. It divides the surface of the earth into 60 zones, and relies on a flat, 2D model.

A **datum** is a set of constant, known points used to model the earth's surface in a particular area, such as North America. Many datums exist because many people all over the world had the same idea to collect and analyze points to create a representative model of the earth.

In North America the most commonly used datum is NAD 83, which is based on 250,000 points across North America. In 2007, NAD 83 was updated and the resulting datum is called NAD 83(NSRS2007). HARN (High Accuracy Reference Network) is a statewide or regional upgrade in accuracy of NAD 83. Another widely used datum is the World Geodetic System of 1984 (WGS 84), which the US Department of Defense uses.

Census shapefiles are already in NAD 83; however, North America is a big place and many GIS projects are focused on a much smaller area, like county level or even neighborhood level. Therefore, it becomes important to apply an appropriate **projection** for a much smaller area. The benefit is twofold. It will give a better visual representation of local geography and it will make distance more accurate.

I sincerely wish there was one quick and easy way to do projections. There just isn't. You will need to apply various projections depending on the location's geographic scale and where on the earth you are mapping.

A GUIDE TO MAP PROJECTIONS

If, instead of just one chapter on projections, you would like to read a whole book about it, you are in luck. A fabulous book called *Lining Up Data in ArcGIS*, second edition, by Margaret Maher (Esri Press 2013) discusses all things related to map projections, including creating custom map projections.

Projecting shapefiles when the geographic coordinate system is known

We downloaded shapefiles from the US Census Bureau. They come in NAD 83. What if you didn't already know this? Or you get shapefiles from another source and don't know anything about their projection. Below are the steps to identify a shapefile's coordinate system, datum, and projection.

1 Get datum information

1. Open a new empty map document.
2. Click the Add Data button, navigate to your save folder, and select the file created in chapter 1 called alabamacounties (yours will likely have a different name), which contains all counties in your state. (If you are unable to find this file, you can access as similar one at C:\EsriPress\GIS20\03.)

   NOTE: Do *not* select the counties file containing all the counties for the United States. That needs a different projection.

3. Right-click the layer name in the table of contents, and then click Properties.

4. Click the Source tab. Notice the Data Source section displays information about the shapefile. Of particular importance here is Datum: D_North_American_1983.

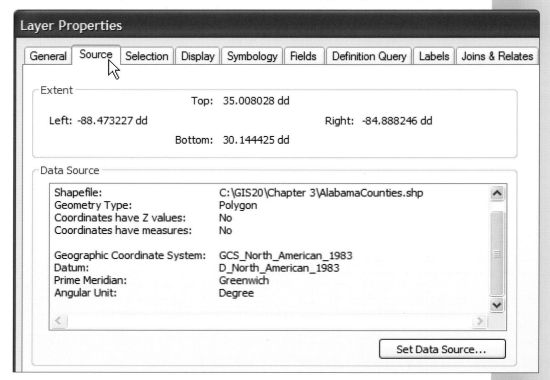

5. Click OK to close the Properties dialog box.

Another place to find datum information is in the data frame properties.

6. Right-click anywhere within the data frame (the window where your map is) and click Data Frame Properties.

> **NOTE:** You must be in data view, not layout view.

7. Select the Coordinate System tab and notice NAD 1983 is highlighted. Click OK to close the box.

Because we need to apply a state-level projection, it is important to figure out which one to use. It varies depending on state.

2 Look up UTM and state plane projections for your state

Next we need to fine-tune the coordinate system of this file to make it a better fit for Alabama (and for your state). At this geographic scale (state level), both UTM and state plane are good choices, but the best choice depends on which state you are mapping. You will select whichever one provides a singular projection for the whole state.

Let's figure out whether UTM or state plane will be best and what the code is for Alabama (and your state).

1. Open an Internet browser and navigate to the State Plane Coordinate System Designations website created by Rick King, **http://home.comcast. net/~rickking04/gis/spc.htm**. This site provides a quick and easy way to determine which projection is best.

2. Scroll down to your state. The first thing to determine is whether more than one FIPS zone exists for your state. The FIPS zone denotes the state plane projection.

Alabama is covered by FIPS zones 0101 and 0102, which relate to the east and west sides of the state, respectively. What this means is for state plane projection, the state is split into two zones. California, on the other hand, is split into six zones and Connecticut only has one zone.

3. Next, notice now how many UTM zones exist for your state. Alabama has one UTM zone, which is 16. This makes UTM a preferable choice for this state, since the whole state falls within one zone. Decide here if you would like to use UTM or state plane. If your state has multiple UTM and state plane zones, see step 3 below. If not, move on to step 4.

(Optional) 3 When a state has multiple UTM and state plane zones

Let's say that for your state, multiple UTM and state plane zones exist. You may be able to use a custom state system located in the State Systems folder under Projected Coordinate Systems. Alaska, California, Georgia, Michigan, Texas, Wisconsin, Florida, Georgia, Mississippi, Virginia, Idaho, and Oregon all have custom state systems. However, if you are mapping a state that doesn't have a custom system listed in the State Systems folder, and has multiple UTM and state plane zones, that leaves you with three options.

Option 1: Select the UTM or state plane that covers the largest portion of the state, and be OK with the knowledge that accuracy may be slightly off for parts of the state outside of the selected zone.

You can use figures 1 and 2 on the following page to help decide which zone to use. For example, Colorado falls *mostly* within UTM 13 (figure 2) and is evenly distributed between state planes 0501, 0502, 0503 (figure 1). In this case, projecting in UTM 13 would be a reasonable choice, assuming that very high distance accuracy in the areas outside of UTM 13 is not a primary concern. If you need a visually correct map with fairly accurate distance for most of the state, this option is for you.

> **NOTE:** If multiple UTM and state plane zones exist, and they all equally cover your state, just pick one.

In order to complete this exercise you must decide now which coordinate system you would like to use.

Option 2: Do an Internet search to see if your state has a custom projection. If so, you can import this system within the Project dialog box.

Option 3: Create a custom coordinate system for your map. You can find information on how to do this in the Help menu under "new coordinate system." (Beginners shouldn't try this unsupervised.)

Figure 1

State Plane Coordinate System Zones, 2004

Due to space constraints, Alaska and Hawaii not shown.

Figure 2

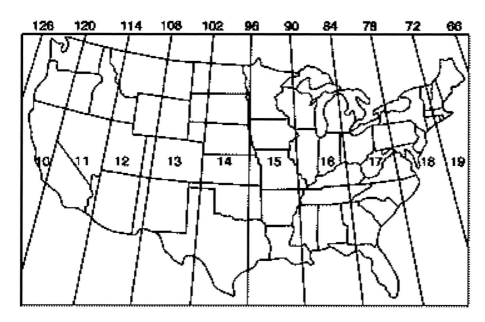

4 Project your file

1. Click the ArcToolbox button 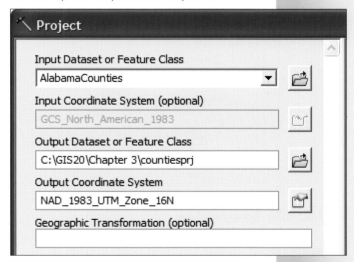 to open ArcToolbox.

2. Expand the Data Management Tools toolbox, then expand the Projections and Transformations toolset, and then expand the Feature toolset.

3. Double-click the Project tool. If we had multiple files, we could use the Batch Project tool to do many files at once.

4. In the first field, click the Input Datasets or Feature Class arrow and select the alabamacounties shapefile.

5. Input Coordinate System will fill in automatically.

6. The Output Dataset or Features Class field represents where to save what will be a newly projected shapefile. Navigate to your save folder and name it **countiesprj** (for counties that have been projected). The Save as type should be Feature classes. Click Save.

   NOTE: If it is unclear how to navigate to the C drive, revisit "Finding Files" in the introduction of this book.

7. Click the Output Coordinate System button.

8. Double-click the Projected Coordinate Systems folder.

9. Double-click the UTM, State Plane, or State Systems folder, depending on which one you have decided to use.

10. For UTM, double-click NAD 83 folder. For state plane, double-click NAD 1983 (US Feet). For State Systems, select the best option for your state (if you have multiple options, go with one that is most current, 1983 over 1927 for example, and feet over meters).

11. For UTM, select NAD 1983 UTM Zone 16N (N stands for the northern hemisphere). For state plane, select the correct one for your state. Click OK to close each window. Close ArcToolbox. →

PROJECTION (.PRJ) FILE

All projection information for a shapefile is stored in a projection (.prj) file as a part of the shapefile.

PROJECTION: TEMPORARY OR PERMANENT

The projection method in this exercise will permanently project the shapefile. If you would prefer to only project the shapefile temporarily for this one project, do the following: right-click within the data frame, click Data Frame Properties, click the Coordinate System tab, and navigate to the coordinate system you would like to use. Double-click the system, click the Change button, and click OK three times. This does not change anything about the original shapefile. It only changes the projection in the data frame.

NEW IN ARCGIS 10.1: SEARCH FOR COORDINATE SYSTEMS

A new tool available in ArcGIS 10.1 provides the ability to search for coordinate systems. You can type in the name of any geography, such as a state's name, click the Search button 🔍, and coordinate systems related to that geography will populate search results.

You can also set a spatial filter and use the Current Visible Extent to do a geographic search based on the map you are looking at in ArcMap. The software can sift through over 4,000 coordinate systems to provide you with those relevant to the geography you have open in front of you. This is very handy indeed. You can also use the Add Coordinate System tools to create new coordinate systems and import others.

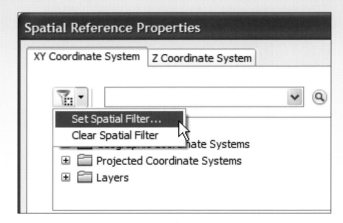

5 Add file to ArcMap

It is necessary to open a new ArcMap session before you are able to see any changes.

1. In the upper left corner go to File > New, and then select Blank Map (if the option is given). Click OK. When prompted to "Save Changes to Untitled?" click No. You should see a blank, empty workspace.

2. Use the Add Data button to add the new projected file countiesprj.shp to ArcMap. Notice how it looks slightly different. In some cases, depending on the shape of the state, it might look very different.

3. In addition to looking at the geography, confirm that it worked correctly by right-clicking the layer name in the table of contents, clicking Properties, and then clicking the Source tab. Notice the projected coordinate system information.

4. Close ArcMap.

Comparison of Coordinate Systems

NAD 83

UTM

North American Datum 1983 is the base coordinate system for many shapefiles, including those from the Census. Leaving the map in this system visually and distance wise slightly skews it. If distance is not important for your project, and visually the state looks okay, you may consider leaving it in NAD 83.

Projecting the file to UTM16 elongates the state and gives more distance accuracy. Depending on the shape of the geography, the visual difference can be significant. Comparing the two scale bars shows the distance differentiation to be about 25 miles.

THE UNITED STATES

For the entire United States, one good option is to use North_America_Albers_Equal_Area_Conic as the projection instead of leaving it in NAD 83. To find this coordinate system, use the new search tools within the Spatial Reference Properties dialog box.

FIVE CLUES YOU HAVE A PROJECTION PROBLEM (and what to do about them)

Problem	Solution
"When I add a shapefile I get an error message that says Unknown Spatial Reference."	This means there is no base coordinate system, such as NAD 83 associated with the file. To add NAD 83 (or another system), do the following: 1. Open ArcToolbox. 2. Expand the Data Management Tools toolset, then Projections and Transformations toolbox. 3. Double-click Define Projection. 4. Use the first drop-down menu to select the shapefile. 5. Click the Coordinate System button. 6. Click the Select button. 7. Click the Geographic Coordinate Systems folder. 8. Select North America. 9. Double-click NAD 1983.prj. 10. Click OK twice.
"I'm working with two shapefiles, but I can only see one of them in my map, even though I see two shapefiles in my table of contents."	This means the shapefiles are not in the same projection. Follow step 1 to check the projection information for each file, decide which is the correct projection, and follow steps 4 and 5 to reproject the files in the same projection.
"The scale bar is obviously wrong. How can I fix it?"	See problem 1 above.
"My map looks skewed. The shape of my state looks kind of strange."	Your map likely needs to have a projected coordinate system like UTM or state plane applied. See steps 1 through 5 in this chapter.
"Everything looks okay but I noticed the coordinate system is undefined. Is this normal?"	No. At a minimum you need to apply a geographic coordinate system like NAD 83, and likely need to reproject to UTM or state plane. See problem 1 above and steps in this chapter.

Preparing data for ArcMap

ArcMap and Microsoft Excel interact very well. That is to say, it is common to prepare data in Excel prior to mapping it. A typical workflow would include preparing data in Excel, instead of trying to do a bunch of complicated manipulation in ArcMap.

The Census Bureau has released 2010 census data and it is a really big deal. Hundreds of data tables are available that provide all sorts of information about the US population. In addition to 2010 files, the Bureau is also committed to releasing annual updates via the American Community Survey.

MAPPING CENSUS 2010: THE GEOGRAPHY OF AMERICAN CHANGE

Mapping Census 2010: The Geography of American Change by Riley Peake (Esri Press 2012) is a very interesting atlas of American change comparing data from the 2010 Census to data from the 2000 Census.

Although we are using a census data table to illustrate data preparation for ArcMap, this exercise is directly applicable to any Excel spreadsheet you would hope to map. Many tips, tricks, and techniques essential to working with Excel spreadsheets in ArcMap are included.

In this exercise, you will download 2010 census data and prepare it for ArcMap. US Census data is raw, messy, real-life data that you will need to prepare for ArcMap, particularly for joining purposes. You will learn about unique IDs and FIPS codes, and how these relate to joining data. You will learn about the wealth of freely available information from the US Census Bureau and useful Excel tips that can enhance your work. Finally, you will create a file to work with in chapters 5 (joining) and 6 (thematic mapping).

Files and tools

Files needed: Files for this exercise will be downloaded from the Internet. Or, if you prefer, you can install this book's DVD and access chapter files at C:\EsriPress\ GIS20\04. Don't know how to install the DVD? See "DVD installation" on page xii.

Tools needed: ArcGIS 10.1 for Desktop, an Internet connection and browser, an unzipping program, and Microsoft Excel (any version).

Downloading data from the US Census

1 Navigate to American Factfinder

1. Go to http://www.census.gov.
2. Hover over the Data link (top, center), and then click American FactFinder. American FactFinder is where all raw data is stored and therefore should be of particular interest to GIS users.
3. Click Advanced Search.

2 Select geography

Our ultimate goal is to map data we are downloading now. Eventually this data will be joined to the countiesprj.shp shapefile, which includes all the counties in our state. First, select the geographic type, which again, is county level.

1. Click the Geographies button on the left. ➔

2. In the "Select a geographic type" list, click County. ➔
3. In the "Select a state" list, click your state.

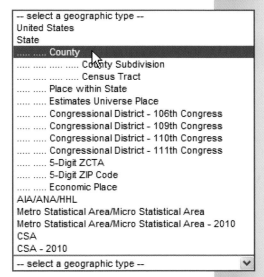

4. In the last list, click "All Counties within <your state>," and then click the Add to Your Selections button. ➔
5. Notice in the upper left corner under Your Selections, this geography has been recorded.
6. Close the dialog box (not the whole Internet browser).

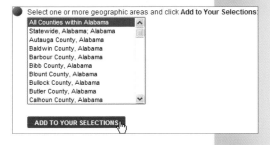

> ## INCREDIBLY USEFUL TIP
> *You are able to select multiple geographies. If you also wanted to download data for cities, for example, you would make that selection here and keep clicking the Add to Your Selections button to continue adding additional geographies.*

3 Search data

You must now choose the data tables to map.

1. In the Topic or Table Name search box at the top of the page, type **Profile of General Population and Housing Characteristics 2010** and click the Go button.

2. Four results are returned. Select the table with the 2010 Demographic Profile SF dataset.

	ID	Table, File or Document Title	Dataset	About
☐	DP-1	Profile of General Population and Housing Characteristics: 2010	2010 American Indian and Alaska Native SF	*i*
☐	DP-1	Profile of General Population and Housing Characteristics: 2010	2010 Demographic Profile SF	*i*
☐	DP-1	Profile of General Population and Housing Characteristics: 2010	2010 SF1 100% Data	*i*
☐	DP-1	Profile of General Population and Housing Characteristics: 2010	2010 SF2 100% Data	*i*

This table provides many key variables from the 2010 Census of Population and Housing.

3. Click the link to explore the types of information included in this data table. Once you are finished looking, click the Back to Search button (upper, left) to go back to the previous screen.

For GIS purposes, data tables must be downloaded from this screen. Downloading it from here provides a file that is organized in a way that will be conducive to working with it in ArcMap, especially for joining and thematic mapping later on.

4 Download data

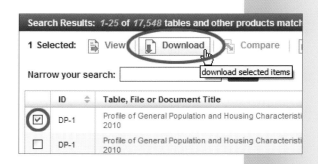

1. Select the box for the desired table, in this case Profile of General Population and Housing Characteristics 2010.

2. Click the Download button to begin downloading the table. →

3. Click the OK for Download box to download a zipped file.

4. Click the Download button to download the file.

5. Navigate to your save folder and save the aff_download.zip file. Unzip the file.

5 Import into Excel

The file we are interested in is DEC_10_DP_DPDP1_with_ann.csv.

Normally, you might simply double-click the file to open it in Excel. There are some peculiarities at work here, though, that have to do with the FIPS code. **Do not simply double-click the file to open it.** You must import it into Excel and change the FIPS code column from numeric to text in order for the FIPS codes to display properly. Also, you cannot double-click the file to open it in Excel and change the column type once you are in either. It will not work. **You must follow these steps.**

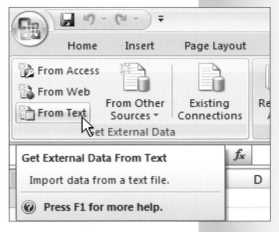

1. Open Excel (these steps use Excel 2007).

2. Click the Data tab, and then click the From Text button. ➔

3. Navigate to your save folder and double-click DEC_10_DP_DPDP1_with_ann.csv.

4. On the Text Import Wizard dialog box, click Delimited as the original data type. Click Next.

5. In step 2 of the wizard, in the Delimiters panel, clear the Tab box and select the Comma box. Click Next. ➔

6. In step 3 of the wizard, in the Data preview panel, select the second column, and then select the Text option in the Column data format panel to change that column type to text. Click Finish and OK. Your data should appear in Excel. ➔

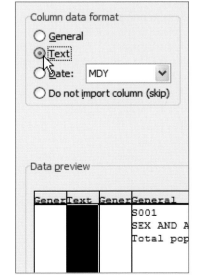

6 Notice the FIPS code

The FIPS code provides a unique ID for every parcel of land in the United States.

States have two-digit codes, counties three, and so a state plus a county code is a five-digit unique identifier for every county in the United States. There are nine states in which the two-digit FIPS code begins with a zero. The entire reason we went through the trouble of importing data into Excel, instead of just opening it, is so we could keep the leading zero on the county five-digit code for these states. If this was not done, nine states would not be able to join.

1. Confirm now that the second column GEO.id2, has five digits. If it does not, the data did not import correctly. Close the table and reimport it following step 5.

Prepping data for ArcMap

The purpose of preparing data in Excel first, instead of in ArcMap, is to simplify the data, focus on what you want to join (and eventually thematically map), and create the optimal spreadsheet to import to ArcMap.

When downloading census data, you get numerous columns of irrelevant information. It is helpful to decide early on exactly what you would ultimately like to display on your map. In chapters 5 and 6, we will join and create a thematic map using senior population. We don't need to bring all of this data into ArcMap because we only need a small portion of it.

7 Clean up the spreadsheet

1. Delete the first column, GEO.id, because it is unnecessary. (Do not delete your precious Geo.id2 column, which is the FIPS code and what we'll use to join this data to a map later on. Seriously, do not delete it.)

2. Keep columns A, B, and C, as they represent the FIPS code, county name, and total population in that county, but remove columns D through AX. Highlight column D by clicking the column heading, and then hold down the Shift key and use the keyboard's right arrow to highlight columns D through AX. Right-click anywhere within the highlighted area and click Delete. This will delete all highlighted columns.

AW	AX	AY	AZ
S024	S024	S025	S025
SEX AND AGE	SEX AND AGE	SEX AND AGE	SEX AND AGE
Total population	Total population	Total population	Total population
62 years and over	62 years and over	65 years and over	65 years and over
HD01	HD02	HD01	HD02
Number	Percent	Number	Percent
8222			12
37780			16.8
4958	18.1	3909	14.2
3690			12.7
10549			14.7
1827			13.5
4242			16.7
21095			14.3
7073			16.7
5832			17.9
7372			13.6
3066			18.2
5081			16.2
2959			17.6
2952			15.8
8987	18	7210	14.4

Context menu overlaying the table:

Calibri · 11 · A A $ · % ,

B *I* ≡ · ▦ · ◇ · A · ⌐₀ ₀⌐ ▦

- ✂ Cut
- 📋 Copy
- 📋 Paste
- Paste Special...
- Insert
- Delete
- Clear Contents
- 🗐 Format Cells...
- Column Width...
- Hide
- Unhide

In this exercise we are mapping the senior population (percentage of population 65 years and older). Lucky for us, the Census Bureau provides this column already. It is contained in column E.

INCREDIBLY USEFUL TIP

Once you have all the columns highlighted, if you simply clicked the Delete button on your keyboard, it deletes the data, but not the columns. Right-clicking within the highlighted area and clicking Delete saves a step.

3. Highlight column F and all columns after F, right-click within the highlighted area, and click Delete. This leaves five columns of data.

	A	B	C	D	E
1			S001	S025	S025
2			SEX AND AGE	SEX AND AGE	SEX AND AGE
3			Total population	Total population	Total population
4				65 years and over	65 years and over
5					
6					
7					
8	GEO.id2	GEO.display-label	HD01	HD01	HD02
9	Id2	Geography	Number	Number	Percent
10	01001	Autauga County, Alabama	54571	6546	12
11	01003	Baldwin County, Alabama	182265	30568	16.8
12	01005	Barbour County, Alabama	27457	3909	14.2

4. Delete all rows at the top of the spreadsheet, leaving only one row for column headers. In order to join this table to a shapefile, we can only have one row at the top with column heading information. Change the column headings to **ID**, **County**, **Pop**, **Seniors**, and **Percent** as shown here.

	A	B	C	D	E
1	ID	County	Pop	Seniors	Percent

AVOID SPACES AND NONALPHANUMERIC CHARACTERS IN COLUMN TITLES

Keep these guidelines in mind when naming columns:

- In order to join, column titles cannot contain any strange characters, nor can they begin with numbers.
- In older software versions, column titles must be 10 characters or fewer to join.
- Column titles should not contain spaces. This is particularly annoying when you accidentally insert a space at the end of a column name. It is impossible to see when looking at the table, yet the file will not join properly.
- Column titles should not contain nonalphanumeric characters such as periods, commas, percent symbols (underscores are OK).

COLUMNS THAT CONTAIN FORMULAS

In ArcGIS 10.1 you are now able to leave formulas in Excel columns; just the numbers are carried over to ArcMap. Before version 10.1, it was necessary to turn formulas into real numbers. To do that in Excel you must go to Copy > Paste Special > Values to copy real numbers over the formulas.

8 Clean up the county column

The County column text reads Autauga County, Alabama, but for labeling purposes later, it would be helpful to just have the county name with no additional text. To accomplish this, use Excel's Find & Replace function.

1. Click the Find & Select button on the Home tab (Excel 2007) (or Find & Replace on the Edit menu in Excel 2000) and select Replace.

2. In the Find what field, type **County, Alabama** (except use your state). There is a space before the word "County." Type it exactly how you see it for your state.

3. You will replace it with no text, so don't type anything in the Replace with field. Click the Replace All button. It should look like the illustration here. →

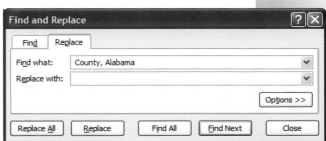

9 Rename the worksheet

Naming Excel worksheets helps you stay organized. Perform the following step to name your worksheet.

1. In the lower left corner double-click the Sheet1 tab (it may be sheet0 depending on the version of Excel you are using) and type **AGE**. You are able to bring in multiple worksheets from the same Excel workbook. →

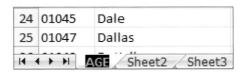

10 Save

1. Click cell A1. Make sure nothing is highlighted.

2. Click the Office button (upper, left) and select Save As. Navigate to your save folder. Rename the file **age.xlsx** (.xls is fine too).

3. Close Excel. You will not be able to open this spreadsheet in ArcMap if it is still open in Excel.

FILE NAMING: AVOID SPACES

Avoid spaces when naming files in ArcGIS. This is kind of a universal rule with file naming in general, but in ArcMap, things can quickly come to a grinding halt because of a space in a file name.

PERCENTAGES

It is common to map a percentage column. You may change the column type to percentage in Excel; however, in ArcMap, this formatting will not be maintained. It is often necessary to change the column type. Here's how to do that:

1. In ArcMap, add the table. Right-click the spreadsheet name in the table of contents, and click Open.

2. Right-click the column heading for the percentage column (or for any column you would like to change the column type), and click Properties.

3. In the Display panel, click the Ellipsis button [...] next to Numeric.

4. In the Category panel, click Percentage. On the right, select "The number represents a fraction. Adjust it to show a percentage." Click the Numeric Options button.

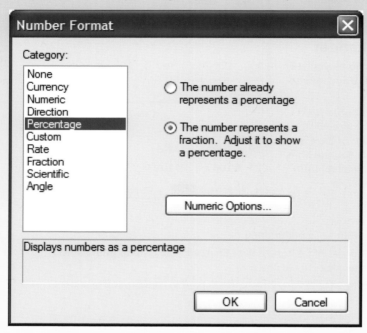

5. Change the number of decimal places from 6 to 2. Click OK three times. Notice how the column has reformatted the column.

These changes are not permanent. You must save the spreadsheet as a part of saving an overall ArcMap project or as a layer (.lyr) file; however, if you use this spreadsheet again in another project, it will be necessary to make these changes again.

You can format the percentage in Excel ahead of time, not by changing the column type, but by formula, where you multiply the ratio by 100 resulting in a percent such as 10.123. But then no % sign is visible in the resulting ArcMap table. The % is helpful to have in the column. When building a legend for the resulting table, if the data contains a % sign, it isn't necessary to type the word "Percent" to indicate the type of data. It will be apparent by the % symbol. Pros and cons exist for both approaches.

CHAPTER 5

KEY CONCEPTS

understanding unique IDs
understanding FIPS codes
joining Excel files to maps

Joining data to maps

One of the most frequently used GIS skills involves connecting an Excel spreadsheet to a shapefile. This is where the magic of GIS happens! Often, the purpose of joining data to a map is to visually display the distribution of a dataset through a thematic map (covered in the next chapter). Joining your own data to a shapefile can be extremely useful.

In this exercise, you will learn how to join your own data to a map, emphasizing the concept of a unique ID.

Files and tools

Files needed: You will need age.xlsx (chapter 4) and countiesprj.shp (chapter 3). Or, if you prefer, you can install this book's DVD and access chapter files at C:\EsriPress\ GIS20\05. Don't know how to install the DVD? See "DVD installation" on page xii.

Tools needed: ArcGIS 10.1 for Desktop.

1 Add two files to a join

1. Open ArcMap. Click the Add Data button and add countiesprj.shp from chapter 3. (If you are unable to find this file, you can access as similar one at C:\EsriPress\ GIS20\05.)

2. Click the Add Data button, add age.xlsx by double-clicking the file name, and then double-click AGE$. If you did not change the name of the worksheet in the last exercise, the existing worksheet will be called Sheet1$ or Sheet0$. Worksheets are denoted with $ in the name.

2 Double-check and find the FIPS columns

1. Check to make sure the data is correct. The AGE$ data table should now appear in the table of contents. To view the data table, right-click the data table name, and then click Open. Review the data to make sure it looks as you would expect it to look.

To join data to maps, we must link two columns that have overlapping data, one column from the data table and its comparable column in the map layer.

2. Identify the two columns you will use for joining by opening the attributes table for each. The table for Age is already open. Right-click the county shapefile in the table of contents, and then click Open Attribute Table. Notice two tabs are now open at the bottom of the table. →

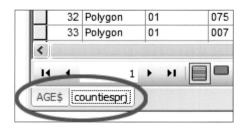

3. Click each tab in the lower left corner, review each, and find two columns that match. The column names do not have to be the same, but the content of the columns does. In this example, the column name in the shapefile attribute table is GEOID and the column name in the spreadsheet is ID.

Table									
FID	Shape *	STATEFP	COUNTYFP	COUNTYNS	GEOID	NAME	NAMELSAD	LSAD	CLASSF
0	Polygon	01	005	00161528	01005	Barbour	Barbour County	06	H1
1	Polygon	01	113	00161583	01113	Russell	Russell County	06	H1
2	Polygon	01	087	00161569	01087	Macon	Macon County	06	H1
3	Polygon	01	097	00161575	01097	Mobile	Mobile County	06	H1
4	Polygon	01	031	00161541	01031	Coffee	Coffee County	06	H1
5	Polygon	01	059	00161555	01059	Franklin	Franklin County	06	H1
6	Polygon	01	115	00164997	01115	St. Clair	St. Clair County	06	H1
7	Polygon	01	099	00161576	01099	Monroe	Monroe County	06	H1
8	Polygon	01	015	00161533	01015	Calhoun	Calhoun County	06	H1
9	Polygon	01	081	00161566	01081	Lee	Lee County	06	H1
10	Polygon	01	069	00161560	01069	Houston	Houston County	06	H1
11	Polygon	01	127	00161589	01127	Walker	Walker County	06	H1
12	Polygon	01	035	00161543	01035	Conecuh	Conecuh County	06	H1
13	Polygon	01	061	00161556	01061	Geneva	Geneva County	06	H1
14	Polygon	01	133	00161592	01133	Winston	Winston County	06	H1
15	Polygon	01	009	00161530	01009	Blount	Blount County	06	H1
16	Polygon	01	083	00161567	01083	Limestone	Limestone County	06	H1
17	Polygon	01	105	00161579	01105	Perry	Perry County	06	H1
18	Polygon	01	011	00161531	01011	Bullock	Bullock County	06	H1
19	Polygon	01	001	00161526	01001	Autauga	Autauga County	06	H1
20	Polygon	01	073	00161562	01073	Jefferson	Jefferson County	06	H1
21	Polygon	01	129	00161590	01129	Washington	Washington County	06	H1
22	Polygon	01	053	00161552	01053	Escambia	Escambia County	06	H1
23	Polygon	01	067	00161559	01067	Henry	Henry County	06	H1
24	Polygon	01	025	00161538	01025	Clarke	Clarke County	06	H1
25	Polygon	01	039	00161545	01039	Covington	Covington County	06	H1
26	Polygon	01	003	00161527	01003	Baldwin	Baldwin County	06	H1
27	Polygon	01	093	00161573	01093	Marion	Marion County	06	H1
28	Polygon	01	055	00161557	01055	Etowah	Etowah County	06	H1

Table				
ID	County	Pop	Seniors	Percent
01001	Autauga	54571	6546	12
01003	Baldwin	182265	30568	16.8
01005	Barbour	27457	3909	14.2
01007	Bibb	22915	2906	12.7
01009	Blount	57322	8439	14.7
01011	Bullock	10914	1469	13.5
01013	Butler	20947	3489	16.7
01015	Calhoun	118572	16990	14.3
01017	Chambers	34215	5706	16.7
01019	Cherokee	25989	4651	17.9
01021	Chilton	43643	5921	13.6
01023	Choctaw	13859	2519	18.2
01025	Clarke	25833	4174	16.2
01027	Clay	13932	2449	17.6
01029	Cleburne	14972	2361	15.8
01031	Coffee	49948	7210	14.4
01033	Colbert	54428	9463	17.4
01035	Conecuh	13228	2362	17.9
01037	Coosa	11539	1970	17.1
01039	Covington	37765	6939	18.4
01041	Crenshaw	13906	2210	15.9
01043	Cullman	80406	12810	15.9
01045	Dale	50251	6759	13.5
01047	Dallas	43820	6165	14.1
01049	DeKalb	71109	9875	13.9
01051	Elmore	79303	9436	11.9
01053	Escambia	38319	5812	15.2

It is imperative you understand the concept here. We have two columns and we are going to link the map to the data table using these columns. They contain identical information. Note the five-digit FIPS code in each. You can even sort these so you can compare line by line.

4. Close the attribute tables.

3 Join the data table to a map

1. In the table of contents, right-click the countiesprj.shp shapefile (not the data table from Excel).

2. Click Joins and Relates, and then click Join.

3. In the "What do you want to join to this layer?" field, select "Join attributes from a table."

4. In the "Choose the field in this layer that the join will be based on" field, select the appropriate column heading, in this case GEOID.

5. In the "Choose the table to join to this layer" field, AGE$ will already be selected.

6. In the "Choose the field in the table to base the join on" field, ID will already be populated. (If it is not, the column is incorrectly formatted. The genesis of this error is not importing the Excel spreadsheet into Excel, but rather just opening it. You will need to close ArcMap and Excel, go back to chapter 4, step 5.)

7. Select the "Keep Only Matching Records" option and click OK. To save a step here, skip the Validate Join button.

> ### INCREDIBLY USEFUL TIP
> *If the join does not work, go back and select Validate Join, which will give you clues about why it didn't work.*

4 Verify the join worked correctly

1. Right-click the shapefile name, and then click Open Attribute Table.

2. Scroll to the far right to see if data from the spreadsheet has been appended to the end of the attribute table. You should not see any error messages or null values. →

ID	County	Pop	Seniors	Percent
01005	Barbour	27457	3909	14.2
01113	Russell	52947	6720	12.7
01087	Macon	21452	3031	14.1
01097	Mobile	412992	53321	12.9
01031	Coffee	49948	7210	14.4

3. Double-check the number of records in the Age$ tab by right-clicking the file and clicking Open. In the lower right corner, the number of records is listed (in this case, 67). Now check the number of records in the newly joined shapefile—it should be the same number. If it is not, then the two columns are not identical and must be corrected.

4. Close the attribute table.

5 Create a new shapefile

When files are joined, it is a temporary join. To permanently join these files, create a new shapefile out of one that was just joined. To do this, do the following:

1. In the table of contents, right-click the shapefile name, click Data, and then click Export Data.

2. Click the Browse button 🗁 to browse to your save folder, name the new shapefile **agejoined** (no spaces in file names). The Save as type should be shapefile. Click Save. Verify it is saving where you would like, and then click OK.

3. When asked if you want to add the exported data to the map as a layer, click Yes. Notice the new file added to the table of contents.

4. The original Excel file and shapefile are no longer needed. To remove them, right-click the AGE$ file and click Remove. Then, right-click the original county file and click Remove.

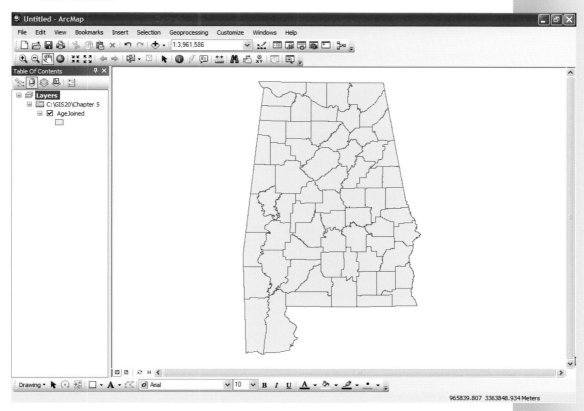

Congratulations! You now have a permanently joined a shapefile (agejoined) that contains data about the senior population. We will use this file in the next chapter.

CHAPTER 6

Creating thematic maps

Thematic mapping enables you to show distribution of data across geography. It is one of the most frequently used GIS tools.

In this exercise, you will create a thematic map with graduated colors emphasizing fundamentals, including color shading, customizing legends, and layouts. You will map the distribution of the senior population by county.

Files and tools

Files needed: You will need agejoined.shp (chapter 5). Or, if you prefer, you can install this book's DVD and access chapter files at C:\EsriPress\GIS20\06. Don't know how to install the DVD? See "DVD installation" on page xii.

Tools needed: ArcGIS 10.1 for Desktop.

1 Create a thematic map

1. The shapefile agejoined.shp should still be opened from the last exercise. If not, use the Add Data button to add this file to ArcMap.

2. In the table of contents, right-click the layer name, and click Properties. Click the Symbology tab.

3. On the left, select Quantities and Graduated Colors.

4. In the Value list, click the column of data you want to map, in this case Percent. Click OK. →

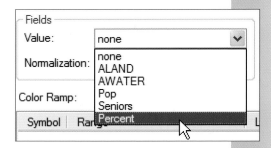

2 Change the color ramp

1. Right-click the layer name in the table of contents, click Properties, and then click the Symbology tab. Select a more suitable color ramp if needed. Click OK.

> ### INCREDIBLY USEFUL TIP
> *Monochromatic blue is a good choice for thematic maps because it looks professional and readers can easily distinguish between the shades of blue. Grays, tans, and greens are also good choices. Avoid reds, oranges, pinks, and purples.*

3 Fix the percentages

The percentages display in an obnoxious way in the table of contents. They would also display this way in the legend. Let's fix this.

1. Right-click the layer name in the table of contents and click Open Attribute Table.

2. Scroll far right and look at the Percent column. The values are displayed here with one decimal place (e.g., 9.1), but in the table of contents the values are displayed with six decimal places (e.g., 9.100000). Right-click the Percent column heading and click Properties.

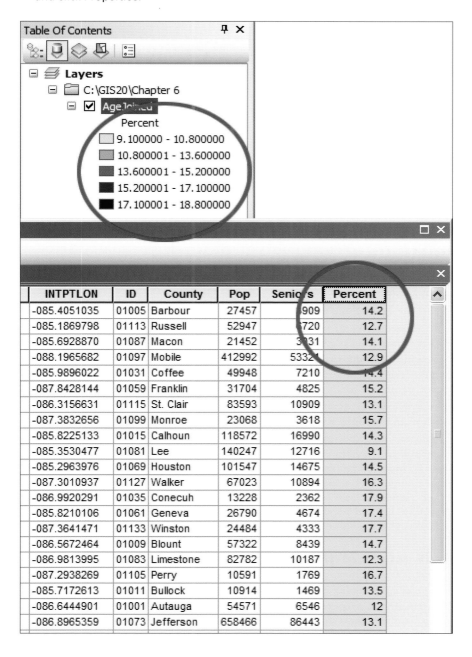

3. Click the Ellipsis button ⬚ to the right of the word Numeric.

4. In the Category panel, click Percentage. Make sure "The number already represents a percentage" is selected, and then click the Numeric Options button.

5. Change the number of decimal places from 6 to 1. Click OK three times. Close the attributes table.

6. Right-click agejoined in the table of contents and click Properties. Select the Percent column from the Value list again. Click OK. This refreshes the table of contents and displays the percentage in a reader-friendly way. ➜

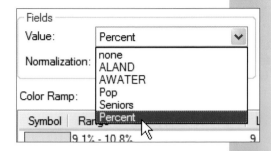

4 Change the way the legend breaks

The natural breaks method for thematic maps is suitable for informal mapping; however, for a more sophisticated approach, you should manually break the legend. Rather than deciding where to break it randomly, use the average for your dataset as the first break point (thereby establishing a baseline for normalcy).

Classes are the data ranges that appear on a map's legend. The default is five. Four intervals may be preferable for a general, nontechnical audience.

1. Right-click the layer name in the table of contents, click Properties, and click the Symbology tab.

2. Change the number of classes from 5 to 4 in the Classes box and click the Classify button. ➜

3. Look in the statistics box (upper, right) to determine the mean. Write down that number. In this example, it is 15.1 percent.

4. To change the first break point to the mean of your data (in this case 15.1), manually type over the given numbers in the Break Values section.

5. For Alabama, the data distribution here only goes to 18.8 percent, so the break points after the 15.1 percent will have to be very small increments. Change the second and third break points to something that makes sense for your data range. Never change the last break point, as this is the natural end of your data. Here, we'll use 16.1 and 17.1 and leave the last break point at 18.8. Click OK twice and look at the map. Notice you can more easily see a pattern emerging from the data. ➜

5 Place the highest values at the top

To create a map that emphasizes the concentration of a variable (versus the lack of a variable), construct the legend so the highest values are at the top. The legend has not been created yet, but the content of the table of contents will be the content of the legend. Sort the classes so the highest values are at the top.

1. In the table of contents, right-click the layer name, click Properties, and click the Symbology tab.

2. Click the Range column heading and choose Reverse Sorting.

3. Click the Symbol column heading and choose Flip symbols.

6 Change the color (again)

Now that you have applied color to your map, a pattern may be emerging. You may be able to make this pattern even clearer. De-emphasize less interesting areas by not shading a color, and emphasize areas of interest by applying color. The less color on your map, the easier it will be to read. Using white to denote the lesser values in the map will make a pattern easier to see and look more professional.

Change the map colors to emphasize areas with high proportions of seniors.

1. Change the lowest range to white (not hollow). You can simply double-click the color and change the fill color. Next, change the second range to a light gray. Change the top two classes with the highest values to medium and dark blue, with dark blue signifying the highest values. Click OK and look at your map.

This symbology choice de-emphasizes low data values, while emphasizing high data values with color.

7 Label the counties

1. Turn on labels for the counties by right-clicking agejoined, clicking Properties, clicking the Label tab, and selecting the "Label features in this layer" box in the upper left corner. Applying a halo will help the label stand out. How to turn on labels is covered in chapter 2.

8 Create a layout

1. On the View menu, click Layout View. Create a layout that includes a legend, title, source, north arrow, and scale bar. For a review of these steps, see chapter 2.

THEMATIC MAP TITLES

When deciding what to call a thematic map, consider beginning the title with the phrase "Distribution of …", for example, "Distribution of Poverty Rates by State, 2010." This phrasing works well with many maps.

9 Fix the legend

Often the legend will be clearer by removing layer names and column headings. In this example, the legend doesn't really need the text "agejoined" and "Percent."

1. To remove this text, double-click the legend to go directly to the Legend Properties.

2. Click the Items tab.

3. There is a secret menu here accessible by right-clicking agejoined located underneath the Select All and Select None buttons. Do this, and then click Properties. →

4. Click the General tab and clear the "Show Layer Name" and "Show Heading" boxes. Click OK twice. →

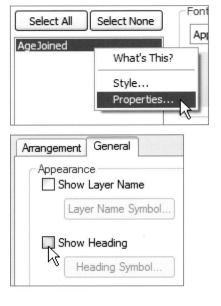

TRANSPARENT LEGEND ELEMENTS

You may encounter an issue with the legend where a see-through section between the legend background color and the box around the legend exists. The Gap tool in Legend Properties might help fix this issue, but it is likely you will need to delete the legend and reinsert it again, this time selecting white as the background color the *first* time through the menu.

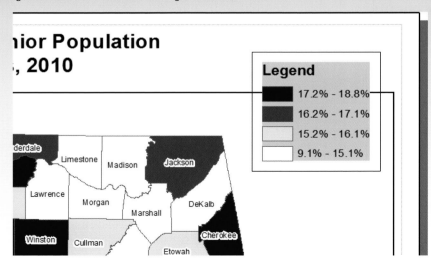

CITING THE CENSUS: APA WEB CITATION

Maps must include information on the sources of data and shapefiles used, and the map creator. Here is one way to cite Census data and shapefiles:

For Data

2010 Census of Population and Housing, Profile of General Population and Housing Characteristics 2010. Retrieved June 28, 2010, from **http://factfinder2.census.gov/ faces/nav/jsf/pages/index.xhtml**.

For shapefiles

US Census Bureau. (2011, Jan1). 2011 TIGER/Line Shapefiles for: Alabama. Retrieved June 28, 2012, from US Census Bureau: **http://www.census.gov/cgi-bin/geo/ shapefiles2011/main**.

10 Save the project

1. On the File menu, click Save As and navigate to your save folder.
2. Name this file **Seniors.mxd** and close ArcMap.

CHAPTER 7

Working with data tables

Manipulating data tables (attribute tables) includes frequently performed tasks such as adding and deleting columns, editing values, and performing calculations. Although the bulk of data manipulation is best achieved outside of ArcMap, it is important to learn the fundamentals of data manipulation within ArcMap. The ability to edit data will greatly strengthen your analysis and improve your maps.

In this exercise, you will edit data tables and work with multiple tables simultaneously.

Files and tools

Files needed: You will need agejoined.shp (chapter 5). Or, if you prefer, you can install this book's DVD and access chapter files at C:\EsriPress\GIS20\07. Don't know how to install the DVD? See "DVD installation on page xii.

Tools needed: ArcGIS 10.1 for Desktop.

1 Add a shapefile and open the attribute table

1. Click the Add Data button to add agejoined.shp created in chapter 5. (If you are unable to find this file, you can access a similar one at C:\EsriPress\GIS20\07.)

2. Open the attribute table by right-clicking the layer name in the table of contents and clicking Open Attribute Table. You should see data in spreadsheet format.

2 Edit existing data in an attribute table

If you try to edit any of the data in the table, you will notice the values cannot be changed. You must first make the table editable.

1. Move the attribute table out of the way. Click the Editor Toolbar button ⚒ and a new Editor toolbar will be visible.

   If you have an older version of the software, go to the View menu, click Toolbars, and then click Editor.

2. Click the Editor button, and then click Start Editing. Notice the top row of the attributes table turns white, which indicates that the table is now editable. →

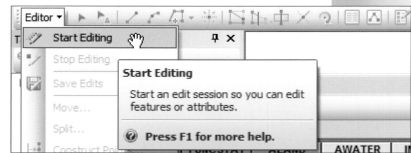

3. Scroll left and find the ClassFP column. Type over any of the values in this column.

4. When you are finished, click Editor again, and then click Stop Editing.

5. Save Edits when prompted and close the attribute table.

3 Edit data outside of the attribute table on a polygon-by-polygon basis

You can also edit data within an information box instead of doing it within the attribute table. One advantage of this method is that you can click individual polygons and edit data for that polygon. To do this, do the following:

1. On the Editor toolbar, click the Editor button, and then click Start Editing.

2. On the Editor toolbar, click the Attributes button ▦.

3. On the map, click any county. Notice a new window opens on the right with all the information from the underlying attributes table. →

4. Click the ClassFP field and replace H1 with **H3** for the value in this field.

5. On the Editor toolbar, click Editor, Save Edits, and Stop Editing.

6. Close the Editor toolbar by right-clicking the toolbar and clicking Editor.

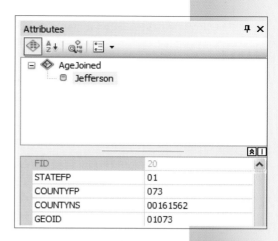

4 Add a column to the attribute table

1. Open the agejoined attribute table. On the Table toolbar, click the Table Options button ▾, and click Add Field. →

2. Type the name of the new column. Call it **NewColumn**. For the Type, select Float. Float is a flexible column type, so it is often used. Float (as well as double) is useful because the number can be in decimal or whole number.

3. Leave the Precision and Scale as zeros for the maximum digits. Click OK.

The column is added to the end of the attribute table.

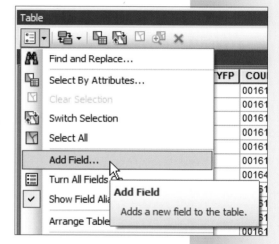

NUMERIC COLUMN TYPES

Generally, you should use float or double (also known as a double float) as the column type. The double float is beneficial because you can have the maximum number of digits for the number, with no rounding (after 15 digits, plain old floats start to round, but doubles do not).

5 Make calculations

Calculations can be performed in or out of an editing session. Let's pretend we have not calculated the percentage of the senior population. For the purpose of this exercise, recalculate the percentage of the senior population.

1. Right-click the NewColumn heading.

2. Click Field Calculator and accept the warning.

3. The formula to calculate the percentage of seniors is "seniors divided by population." Double-click Seniors from the list of variables, click the / button (the division symbol), and then double-click Pop (Population) from the variables list. The formula is automatically filled in as you go. Click OK. ➡

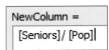

NewColumn =

[Seniors] / [Pop]|

The software calculates the percentage, but the column represents a fraction; we would like it to display as a percentage. Reformat the column.

4. Right-click the column heading and click Properties. Click the ellipsis […] to change the column type.

5. Select Percentage on the left, and then select the second option on the right, "The number represents a fraction. Adjust it to show as a percentage."

6. Click the Numeric Options button. Change the decimal places from 6 to 1.

7. Click OK three times.

8. Sort the column to see which counties had the highest percentages of seniors. Right-click the column heading and click Sort Descending.

6 Delete columns

You may notice several prebuilt columns when you download shapefiles from the census. Some columns are useful for understanding what geography is within the shapefile, but many of these columns can be deleted because they are rarely used. You can delete columns two ways. The method you use depends on how many columns of data you want to delete. If it is only one or a few columns, you would use the following method (for several columns, a different method must be employed).

To delete one column of data, do the following:

1. Right-click the LSAD column heading and click Delete Field.

2. Select Yes to delete the LSAD column.

To delete multiple columns of data, do the following:

3. Close the attribute table.

4. In the table of contents, right-click the layer name, click Properties, and then click the Fields tab (center).

5. Click the Turn All Fields Off button ⋮ and notice that the check boxes next to all the columns are cleared.

6. Now, select only those columns you want to keep. The essential columns to keep are GEOID, NAME, and Percent. Click OK.

7. In the table of contents, right-click the layer name and open the attribute table. Notice that now only three columns are displayed. The other columns are still there but hidden. Just hiding the columns might work for some projects, but for many other projects you really might only want these three columns.

8. Close the attributes table.

9. Let's essentially make a copy of this file, and in so doing, capture only these three columns. Right-click the layer name in the table of contents, click Data, and then click Export Data.

10. Click the Browse button 📤 and navigate to your save folder. Type a new name such as **Minifile**, make sure the Save as type is Shapefile, and click Save.

11. Click OK and Yes to add the file to the map as a layer.

12. Open the attributes table to verify you have only the columns you exported (plus a couple of others that are standard).

7 Work with multiple attribute tables

1. Open both attribute tables (agejoined and Minifile).

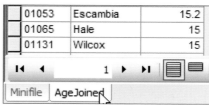

2. Click each tab in the lower left corner of the table's window and notice how you can switch between tables. The tabs can be moved by dragging and dropping. ➔

3. You can also dock the tables to create a split screen. Click and drag the Minifile tab to the center of the table window. A blue circle with four arrows will appear. Drop the tab on the right arrow. The screen will split with one table on each side. ➔

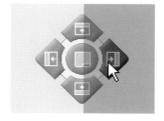

4. To undo the split screen, simply drag the table back into the tab position. Close ArcMap.

CHAPTER 8

KEY CONCEPTS

using address locators

*auto and manual
geocoding*

*using street networks and
other layers*

Address mapping

Address mapping, also called geocoding, is like
creating a pushpin map of addresses. Geocoding is
a skill everyone is likely to need in their GIS work.
You might use this skill for mapping things like
diseases, crimes, client addresses, service addresses,
or anything with a physical location—an address—you
would like to show on a map.

In this exercise, you will learn to geocode by mapping social service agency addresses.

Files and tools

Files needed: All files for this chapter must be accessed from C:\EsriPress\GIS20\08.

- An Excel spreadsheet containing social services in Bexar County, Texas. The file is called Agencies.xls and is located in the chapter 8 folder.
- A street shapefile for Bexar County, Texas. The file is called tl_2011_48029_ edges.shp.
- A water shapefile called tl_2011_48029_areawater.shp.

Don't know how to install the DVD? See "DVD installation" on page xii.

Tools needed: ArcGIS 10.1 for Desktop and an Internet connection.

1 Add the street network shapefile

1. Open a blank map in ArcMap. Click the Add Data button to add the shapefile tl_2011_48029_edges.shp, which was downloaded from the Census Bureau, to your map. These are streets in Bexar County, Texas, where San Antonio is located. This file as been reprojected in state plane.

MORE ABOUT THE STREET NETWORK

The street network file is referred to as the "all lines" file, as well as the "edges" file, on the US Census Bureau website. To download the street network for any county, download the all lines shapefile from the Census website. The file contains all streets in a county, water in linear format, railroads, and trolleys. This file includes anything that is represented as a line.

The MTFCC column contains a key to what each feature represents. Appendix F of the TIGER/Line technical documentation contains the key to the MTFCC.

---

2 Add the list of social services

1. We have some addresses to geocode. Click the Add Data button to add the file Agencies.xls, Organizations$ worksheet.
2. Right-click Organizations$ in the table of contents and click Open. Have a look around. We would like to display 1,144 social service agencies as dots on a map.
3. Close the table.

3 Geocode addresses

NOTE: You must have an Internet connection to complete this exercise.

1. In the table of contents, right-click the Organizations$ file name and click Geocode Addresses. ➡

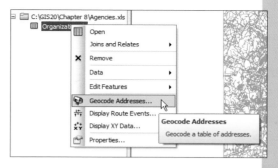

2. In the "Choose an Address Locator to use" window, click "10.0 North America Geocode Service (ArcGIS Online)" and click OK. ➡

3. In the Address field, change <None> to Address, which indicates which column contains addresses.

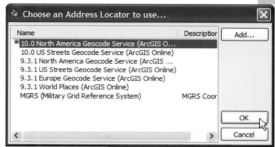

These next steps are a little tricky. The Output Shapefile or Feature Class field indicates where you would like to save what will be the new geocoded file.

4. Click the Browse button 📂 and navigate to your save folder. First you must change the Save as type to Shapefile, instead of File and Personal Geodatabase feature classes.

5. Next, use the drop-down at the top to navigate to your save folder. You can leave the name as Geocoding_Result.shp. Click Save.

6. Click the Advanced Geometry Options button [Advanced Geometry Options...] and click the second option, "Use the map's spatial reference." Click OK. ➡

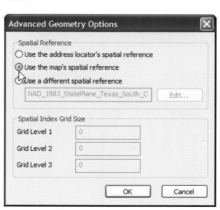

7. Click the Geocoding Options button [Geocoding Options...]. Clear the "Match if Candidates Tie" box. A tied candidate means there are at least two possibilities for one address. The tie is originated based on the match score. ArcMap will place a dot on the map to show the first choice, which is not necessarily the best choice.

8. Click OK twice. Geocoding will begin.

9. Once the geocoding is complete, notice the statistics box. There were 1,132 geocoded, 11 tied, and 0 unmatched. Click the Close button. (See "Composite address locators" later in this chapter.)

10. Have a look at the map. Zoom in and use the Identify tool to get a better sense of what these dots represent.

4 Review the geocoded attribute table

The attribute table of a geocoded file contains a lot of information.

1. Right-click the geocoded file in the table of contents and click Open Attribute Table.

The Loc_Name column indicates what type of address locator was used to match the address. The Status column indicates if it was matched (M), unmatched (U), or tied (T). The Match_addr column shows to what location the given address ultimately was geocoded. ➡

Match_addr
10 Lynn Batts, San Antonio, TX, 78218
100 Citibank Dr, San Antonio, TX, 78245
10002 Cedarbend Dr, San Antonio, TX, 78245
Portland, OR
10004 Wurzbach Rd, San Antonio, TX, 78230
78217
10010 San Pedro Ave, San Antonio, TX, 78216
10026 Trout Ridge Dr, Converse, TX, 78109
1003 Vera Cruz, San Antonio, TX, 78207
10034 Aztec Vlg, San Antonio, TX, 78245

INCREDIBLY USEFUL TIP

If you prefer, you can export the unmatched addresses to Excel and fix them there. Sort the Status column descending with the "U" values at the top. Highlight the records to export, click the Table Options button, and then select Export. Export the records as a dBase table (.dbf), fix them in Excel, and recode them.

COMPOSITE ADDRESS LOCATORS: A LOVE-HATE RELATIONSHIP

We used a composite address locator, 10.0 North America Geocode Service (ArcGIS Online), which makes geocoding easy. We got zero unmatched! Except here's the problem with that: The fourth record had an address of 10002 Weybridge Portland, Oregon. There is no such address. This was a trick. 10002 Weybridge is in San Antonio but was erroneously assigned to Portland. It had the wrong city.

This exemplifies how composite address locators work. Because the address was not found, and because we selected a composite address locator, it placed a point at the center of Portland, Oregon, to represent this address. Notice in the Loc_name column, US_CityState is indicated as the location type, and in the Match_addr column only Portland, OR is indicated. It was not able to find the address, so it did the next best thing and assigned it to Portland. Had there been a ZIP Code, it would have assigned it to the center of that ZIP Code.

Composite address locators are very helpful if you don't mind that if an address cannot be found, it will assign it to the next higher geography such as ZIP Code, city, or state.

If you do not want to use a composite address locator and would rather flag unmatched addresses as unmatched, use the second address locator provided, 10.0 US Streets Geocode Service (ArcGIS Online). This will only match based on address. When this same file is geocoded using that locator, 37 are unmatched and 11 are tied for a total of 48 unmatched.

5 Geocode manually (fixing addresses that did not geocode)

1. In the table of contents, click the layer name of the geocoded file (geocoding_result.shp), click Data, and then click Review/Rematch Addresses.

2. In the Show Results list, click Matched Addresses with Candidates Tied. →

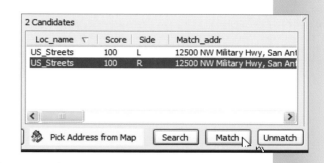

3. For the first and second records, 12500 Northwest Military Highway (Jewish Community Center), the locator doesn't know if it should be matched to the left or right sides of the street. Both candidates were assigned a match score of 100. An Internet search shows an aerial map with the Community Center located on the right side of the street.

4. Select the second candidate, which will place the address on the right side, and click the Match button. →

5. For the next address, 1420 3rd Street, hold down the Shift key and highlight all candidates. Doing this will display these options on the map. Right-click any of the candidates and click Flash to see that particular candidate flash on the map. →

6. Scroll far right and determine the agency's name. The record is for The American Red Cross, 1420 3rd Street, San Antonio, Texas, 78295. An Internet search determines that there is no such location. This is not the address of any Red Cross in San Antonio, and no business is located at this address. The address is incorrect. For this one, do not match it.

For this exercise, we're not going to continue to manually match the addresses, but in real life you might.

7. Click the Close button to close the Interactive Rematch box.

8. Close the Geocoding toolbar by right-clicking the toolbar and clicking Geocoding.

GLOBAL SEARCH AND REPLACE IN ARCMAP

Let's say you have the same street name misspelled throughout your list of addresses. Instead of correcting each one manually, you can make the attribute table editable (see chapter 7) and use the Find/Replace tool to find and replace the misspelled text.

6 Symbolize other layers

Streets and water are two common files you might use with a geocoded file. Now that most of the addresses are geocoded, we should pay attention to symbolizing the street network appropriately, as well as adding water.

1. Right-click the street network in the table of contents, click Properties, and click the Symbology tab.

2. Click the symbol swatch. Select Residential Street from the styles on the left, and change the color to light gray. Click OK twice to close the open windows.

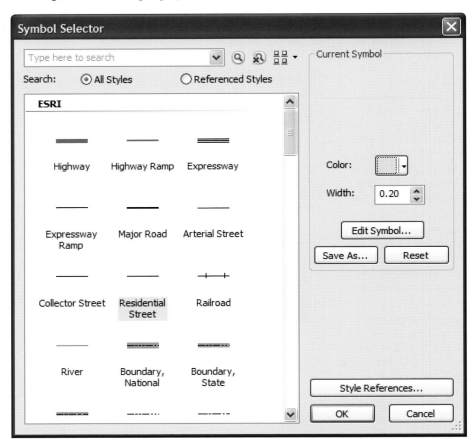

3. Add the tl_2011_48029_areawater.shp. Change the fill and outline color to a dark blue. The outline around water should be the same color as the water itself.

4. In the table of contents, click the List by Drawing Order button.

5. Pull the water layer into first position in the table of contents.

7 Change symbols

Hundreds of symbols are available in ArcMap. Let's change the geocoded symbols to another shape.

1. Right-click the geocoded file in the table of contents, click Properties, and then click the Symbology tab.

2. Click the symbol swatch and notice all the symbols on left side.

3. Click the Style References button, select Crime Analysis, and click OK. Notice that many new interesting symbols have been added to the Symbol Selector. Feel free to look at other symbol palettes. (Bonus exercise 3 illustrates how to import custom symbols.)

These are social service agencies, so a simple circle like Circle 2 will work nicely. →

4. Click Circle 2. On the right, change the size to 7. Click OK twice.

8 Save the project

1. On the File menu, click Save As and navigate to your save folder.

2. Name the project **agencies.mxd**. Close ArcMap.

Circle 2

CHAPTER 9

Creating a categorical map

Categorical mapping is similar to thematic mapping in that color shading is used to indicate values; however, values in a categorical map represent categories instead of numbers. This technique can be used with polygons, for example, to map land-use zoning categories (residential, commercial, and industrial). It can also be used with point data, such as a geocoded file, to map such things as crime (burglaries, assaults). Categorical mapping also works with line data to map different types of streets (residential, major arterial, and highway).

In this exercise, you will categorically map agencies by type of service, learn to understand when categorical mapping is useful, and learn about different symbols.

Files and tools

Files needed: You will need agencies.mxd (chapter 8), tl_2011_48029_areawater.shp (chapter 8), and tl_2011_48029_edges.shp (chapter 8). Or, if you prefer, you can install this book's DVD and access chapter files at C:\EsriPress\GIS20\09. Don't know how to install the DVD? See "DVD installation" on page xii.

Tools needed: ArcGIS 10.1 for Desktop.

1 Review attributes

1. Open ArcMap. On the File menu, click Open, and then navigate to your save folder. Select agencies.mxd. (If you are unable to find this file, you can access as similar one at C:\EsriPress\GIS20\09.)

2. In the table of contents, right-click the geocoded file name and click Open Attribute Table.

3. Scroll far right and review the Type column. This column indicates the type of service each organization provides. Several types of services are provided. This is the type of information (names of things) that lends itself to categorical mapping.

4. Close the attributes table.

2 Create a categorical map

1. In the table of contents, right-click the geocoded layer name, click Properties, and click the Symbology tab.

2. From the Show list on the left, click Categories, and then click Unique values. In the Value field select the column of data to map, which is Type.

The symbol can be changed to anything you like. For this exercise, a smooth circle is preferable to the default dot. To change the symbol, do the following:

3. Double-click the dot next to the check box. From the Symbol Selector, choose Circle 1 (you may need to scroll down depending on what symbol palettes were selected in the last chapter) and change the size from 18.00 to 9. Click OK once. →

4. Click the Add All Values button and click OK to see colored dots displaying data.

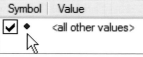

The map will look like someone threw confetti all over it. With this many categories and colors it is hard to read the map clearly. Select only a few types of services to display.

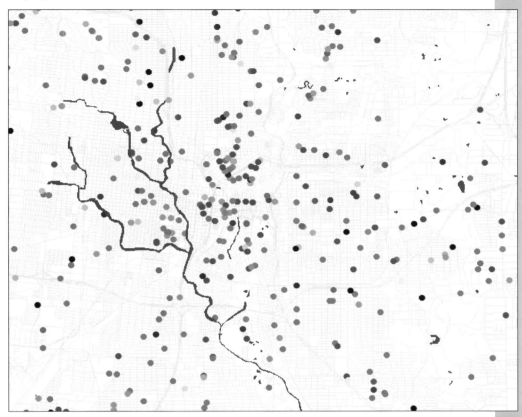

3 Display only certain types of agencies

1. In the table of contents, right-click the geocoded file name, click Properties, and click the Symbology tab.

2. Click the Remove All button to get rid of all the selections.

3. Click the Add Values button.

4. Click the Complete List button. When many items are on the list, ArcMap shortens the list. If the Complete List button is available, click it to ensure you are seeing all list items. →

Add Values

Select the value(s) to add:

ADOPTION SERVICES
ADVOCACY GROUP
AGED HOME
ASSOCIATION FOR THE...
CHILD DAY CARE SERV...
CHILD GUIDANCE AGE...
CHILD RELATED SOCIA...
COMMUNITY CENTER
COMMUNITY SERVICE

OK

Cancel

Complete List

New Value
Lets you add a value to the list above.

Add to List

5. Select the following five types of service: Adult Day Care Centers, Aged Home, Geriatric Residential Care, Geriatric Social Service, and Senior Citizens' Centers. (Hold down Ctrl key to select more than one item at one time, or just select items individually and keep adding.) Click OK.

6. Clear the <all other values> box.

7. Double-click each dot and change the dot color to something that will stand out. Click OK.

4 Group several categories into one category

You can assign multiple categories to one all-encompassing category such as "Senior Services."

1. In the table of contents, right-click the shapefile name, click Properties, and click the Symbology tab.

2. Highlight all five categories by holding down the Ctrl key as you click each category name.

3. Right-click within the dark blue highlighted area and click Group Values.

4. All five categories will now be grouped into one category (the name at this point doesn't matter). Click OK.

5. In the table of contents, click once on the grouped layer to activate the text. Rename this grouping **All Senior Services**.

6. (Optional) To label some of these dots, use the Label tool, which can be found on the Draw toolbar. You can access the Draw toolbar by going to the Customize menu, clicking Toolbars, and clicking Draw. After the tool is activated, expand the capital A. Click the Label tool to activate it, and then click a dot on the map to label it.

7. (Optional) Callout boxes around labels may be helpful. To apply these, with the default pointer, select all labels (hold down the Ctrl key to select more than one). After they are selected, right-click any one of the labels and click Properties. Click the Change Symbol button, and then click the Edit Symbol button. Click the Advanced Text tab and select the Text Background box. Click the Properties button underneath it. Select the second balloon callout, and click OK four times. You may need to reposition the labels to make the anchor visible.

8. Close ArcMap.

DIFFERENT CITIES, DIFFERENT COLORS

Categorical mapping is useful for assigning different colors to different cities, counties, states, or ZIP Codes.

CHAPTER 10

KEY CONCEPTS

*understanding latitude/
   longitude*
using GPS units
creating new shapefiles

GPS point mapping

In addition to mapping a table of addresses (see chapter 8), you can also map a table of latitude and longitude (x,y) coordinates collected in the field using a Global Positioning System (GPS) unit. Latitude and longitude points are frequently used in GIS, for example when a particular place doesn't have an address but you still need to display it as a dot on the map.

In this exercise, you will learn to place latitude and longitude points on a map.

Files and tools

Files needed: All files for this chapter must be accessed from C:\EsriPress\GIS20\10.

- An Excel spreadsheet called XYData.xls containing latitude and longitude points for social service agencies in Bexar County, Texas.
- A street shapefile for Bexar County, Texas. The file is called tl_2011_48029_ edges.shp.

Don't know how to install the DVD? See "DVD installation" on page xii.

Tools needed: ArcGIS 10.1 for Desktop.

1 Add latitude/longitude points

1. Click the Add Data button and navigate to chapter 10 of the EsriPress folder created from this book's accompanying DVD. Double-click XYData.xls and add the desired worksheet XYs$.

2. In the table of contents, right-click the file and click Open. Have a quick look at the data table to confirm you have one X (longitude) column and one Y (latitude) column (as well as a couple of other columns). Close the data table.

2 Place points on your map

1. In the table of contents, right-click XYs$ and click Display X,Y Data.

2. The X and Y fields automatically fill in based on column headers of X and Y. There is no Z field.

3. The coordinate system must be defined. Click the Edit button. Expand the Geographic Coordinate System folder, expand the North America folder, and then click NAD 1983. Click OK twice.

4. When you get a message stating "Table does not have Object-ID field," click OK. It will open with no problems.

> ### *INCREDIBLY USEFUL TIP*
> *Here we selected NAD 83 (for more on projections see chapter 3). NAD 83 is a very common choice. If it doesn't work for your data, check your GPS to determine the geographic coordinate system used.*

3 Create a new shapefile of XY Events

1. Right now the X,Y layer is not saved; it is only a temporary file. To create a new, permanent shapefile of these X,Y points, in the table of contents, right-click the XYs$ Events layer. Click Data, click Export Data, and save the new shapefile. You may need to reproject the shapefile to work with other layers.

2. Close ArcMap.

GETTING LATITUDE/LONGITUDE POINTS OUT OF YOUR GPS UNIT AND INTO EXCEL

Every GPS unit is different, so it is difficult to give specific advice about how to get points from your GPS unit to Excel. One way to do this is to use a new tool in ArcGIS called GPX to Features that enables you to quickly upload those points to ArcMap. If your GPS does not use the GPX file format you must export points out of your GPS unit (perhaps as a text, comma separated value, or database file), open them in Excel, do some minor cleanup work, and save them.

1. Click points with your GPS. These are sometimes called waypoints or points of interest (POIs).

2. Connect the GPS to your computer with a USB cord.

3. Your GPS likely came with software that must be installed on your computer ahead of time. If it has been installed, you can navigate to the folder where your clicked points are stored in your GPS. Often each GPS unit uses a proprietary file format, so don't be surprised if you don't recognize the file extension of your points file.

5. You should be able to open the file in Notepad to make sure it is the correct file. You might try double-clicking the file, and when prompted for which software to use to open it, select Notepad. Save as a text file (.txt). Close the file.

6. Open your text file in Excel. This process will vary depending on what version of Excel you are using. You may need to import the file into Excel, or you may simply be able to open the file.

7. Once your data is imported into Excel, identify your X (longitude or easting) and Y (latitude or northing) coordinate fields, and any corresponding attribute fields. If needed, add or modify the header row to give the columns easy-to-understand titles. Longitude should be titled X and latitude should be titled Y.

8. Save as an Excel file and name the worksheet. Now the file is ready to open in ArcMap.

NEW IN ARCGIS 10.1: GEOTAGGED POINTS TO PHOTO TOOL

The Geotagged Points to Photo tool (in ArcToolbox > Data Management Tools > Photos) reads the X,Y data from photo files (.jpg and .tif only) taken with a digital camera or smartphone. Assuming the camera has the technology to grab the latitude and longitude of the place where the photo was taken, you can quickly and easily upload these points. And here's the cool thing—you can then click on the points in ArcMap to see the photo!

CHAPTER 11

Editing

It may be necessary to change the physical boundary of an existing polygon. For example, if your agency uses a target area boundary to deliver services and would like to extend the service area, editing the boundary would become necessary. Target area boundaries might include things like school districts, voting wards, or neighborhoods.

In this exercise, you will learn common editing tasks such as changing a boundary outline, merging polygons, creating shapefiles out of selected polygons, and appending shapefiles to each other.

Files and tools

Files needed: All files for this exercise, including states.shp, are accessible at C:\EsriPress\GIS20\11. Don't know how to install the DVD? See "DVD installation" on page xii.

Tools needed: ArcGIS 10.1 for Desktop.

1 Open a shapefile and turn on the Editor toolbar

1. Click the Add Data button to add states.shp, located in chapter 11 in the EsriPress folder installed with this book's accompanying DVD.

2. Use the Zoom In tool to zoom in closer to Oregon (or whichever state you want) so you can really see the outline well.

3. Click the Editor Toolbar button ✎. The Editor toolbar will become visible.

The Editor Toolbar button should already be visible because it is a part of the default ArcMap interface. If you have an older version of the software, go to the View menu, click Toolbars, and click Editor.

2 Edit the state outline

1. Click the Editor button Editor ▾ and select Start Editing. This makes not only the polygon boundaries editable, but also the attributes table.

2. A little arrow appears and serves as a pointer. With this arrow, double-click the polygon to edit, in this case Oregon, and notice how several little dots (called vertices) appear in green. Zoom in very close so you can see them. By moving these dots, you can reshape the boundary of the state. Use the Zoom In tool to zoom in close (or the mouse scroll wheel) in order to see the dots clearly. You may need to re-activate the Edit tool ▸ after using the Zoom In tool.

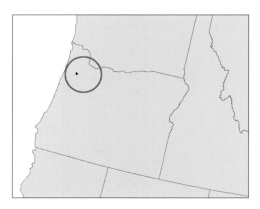

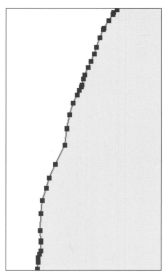

3. Click any one of the dots and drag it to a different position to begin reshaping. Try this with a few more vertices. When you are finished, click anywhere outside the state boundary to complete the edits. ➜

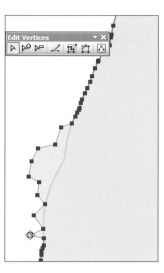

3 Move polygons and cut

One last thing to know is that you can move an entire polygon at once.

1. Zoom out to see multiple states. Click the Edit tool again and click Oregon. Notice the state becomes highlighted in light blue. This means the entire polygon is selected and you can move that polygon. Move Oregon to the left so it doesn't touch the United States anymore.

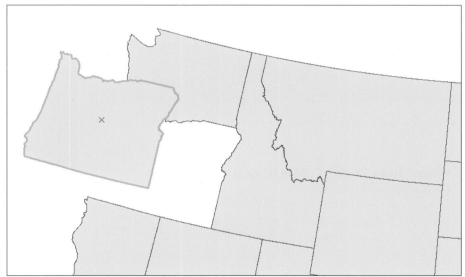

The Editor toolbar contains additional tools. One you may find handy is the Cut Polygons tool 🔲.

2. With Oregon still highlighted and not touching the United States, click the Cut Polygons tool to activate it.

3. Click the left side of Oregon and draw a line across Oregon to the right. On the right side, double-click the Oregon border. Notice, the state has been cut in two. Click the Edit tool again and click on the lower portion of Oregon. Drag that portion down, in order to be able to see the split.

4. Click the Editor tool again and click Stop Editing. You do not need to save your changes.

4 Create new shapefiles out of existing ones

1. Click the Select Features by Rectangle tool. ➡

2. Hold down the Shift key and click on each state to select it (California, Nevada, Utah, and Arizona). They will be highlighted in bright blue.

3. In the table of contents, right-click states.shp.

4. Click Data, and then click Export Data. Type a file name and navigate to your save folder. The Save as type should be Shapefile. Click Save and OK.

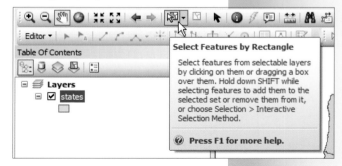

5. Select Yes to add the data as a layer. This will add a new layer to ArcMap.

6. To get rid of the blue outline, from the Selection menu (top, center), click Clear Selected Features. Or you can click the Clear Selected Features button on the Tools toolbar. Notice if you clear the box next to states in the table of contents you can see just the new shapefile.

7. Right-click the new shapefile and remove it, leaving only the states shapefile. Turn the states layer on.

5 Create new features on an existing shapefile

Let's pretend someone has said you can create a brand new state. Of course you would begin statehood by drawing a boundary for the new state.

1. Click the Editor button, and then click Start Editing to make the shapefile editable.

2. Click the Create Features button 📝 at the end of the Editor toolbar.

3. Click states in the first box, and select Polygon as the construction tool. →

4. Move the Create Features box slightly out of the way so you can see the map.

5. Begin drawing a new state next to California. Double-click the polygon when you are finished and you will see a new shape.

The new state boundary may not be a perfect fit with the California boundary. This is a great time to explore some of the other editing tools.

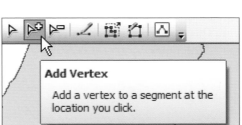

6. Click the Add Vertex button to add additional vertices along the edge that borders California in order to modify it and make it a better fit. →

INCREDIBLY USEFUL TIP

The attribute table for states is also now editable, and you can associate information with this new feature.

7. Stop editing. You do not need to save your edits.

6 Create a new shapefile

Occasionally it may be necessary to create a new shapefile out of thin air.

1. Click the ArcToolbox button 🔴 to open ArcToolbox.

2. Expand the Data Management Tools toolbox, and then expand the Feature Class toolset.

3. Double-click the Create Feature Class tool.

4. In the Feature Class Location field, click the Browse button and navigate to the folder where you will save what will be the newly created shapefile. Click the folder, then click Add.

5. In the Feature Class Name field, type **newsites** (avoid spaces in file names).

6. In the Geometry Type field, select Point. The layer we'll create in a few steps will be composed of points, not polygons.

7. Do not select anything for the Template Feature Class field (see "New Create Features templates" later in this chapter).

8. Next to the Coordinate Systems field, click the Coordinate System button.

9. Expand the Projected Coordinate Systems folder.

10. Select Continental > North America > North America Albers Equal Area Conic. Click OK twice. →

11. Newsites.shp should be added to the table of contents. This is an empty shapefile.

Create Feature Class

Feature Class Location

C:\GIS20\Chapter 11

Feature Class Name

newsites

Geometry Type (optional)

POINT

Template Feature Class (optional)

Has M (optional)

DISABLED

Has Z (optional)

DISABLED

Coordinate System (optional)

North_America_Albers_Equal_Area_Conic

NEW CREATE FEATURES TEMPLATES

In ArcGIS 10.1, "Create Features template" options provide a new way to transfer the structure of an existing attribute table.

7 Construct a new shapefile

1. Click the Editor Toolbar button and Start Editing.

2. Click the Create Features tool and notice a new Create Features box is added to the right.

We are going to add new points to the blank shapefile.

3. In the Create Features box, there will be two files indicated, newsites and states. Click newsites and notice what tools appear under Construction Tools at the bottom. Click states and notice how the tools change, depending on the shapefile's geometry. Click the newsites file name in the Create Features box, and select the Point tool.

4. Click anywhere in Texas to place a new point. Now place two other points, one in California and one in Florida.

5. Close the Create Features box to get it out of the way.

We are not finished yet. We need to add names for these new sites in the attribute table.

6. Right-click the newsites layer name in the table of contents, and then click Open Attribute Table.

Notice no useful options for inputting names for each of these new sites exists. Let's create a column for site names. In order to add a column to the attribute table, the Editor must be turned off.

7. Close the attribute table. Click the Editor button and Stop Editing. Select Yes when prompted to save edits.

Now we are able to add a column to the attributes table.

8. Right-click the newsites layer name in the table of contents, and then click Open Attribute Table.

9. From the Table Option menu (the first button), click Add Field. →

10. For Name, type **Name**, as this will be the name of the new sites. For the Type, select Text, as this column will only contain text.

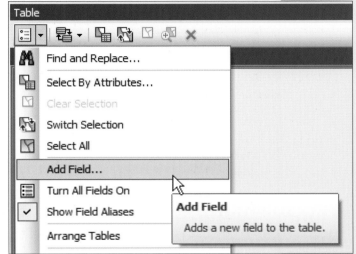

INCREDIBLY USEFUL TIP
If this column were to be a numeric column, float would be a good choice.

11. Leave the length as 50, and click OK. Notice a new column has been added. Close the attribute table.

12. Click the Editor button and Start Editing to make this layer editable.

13. Click the Attributes Table button ▦ on the Editor toolbar. Notice a box appears for attributes.

14. Click the first point in Texas, and type **Texas site** in the Name field in the Attributes panel. →

15. Click the California and Florida sites and do the same for those.

16. When you are finished editing the attributes, click the Editor button, Save Edits, and Stop Editing. Close the Attributes box.

17. Close ArcMap. You do not need to save the project. We will not use it again.

CHAPTER 12

KEY CONCEPTS

writing and erasing queries
creating new shapefiles
based on query results

Creating attribute queries

The ability to query data in an attributes table separates GIS software from graphic design software. Having the ability to query selections of data allows users to analyze communities and problems in more sophisticated ways, to use the software to ask questions, and to perform intelligent analysis based on the answers.

The two most frequently used queries are attribute and location queries (graphics queries also exist and are used least often). An attribute query is simply a data query. Location queries are geographic queries and are covered in the next chapter.

In this exercise, you will create an attribute query that identifies counties with senior citizen population greater than 15 percent.

Files and tools

Files needed: You will need agejoined.shp (chapter 5). Or, if you prefer, you can install this book's DVD and access chapter files at C:\EsriPress\GIS20\12. Don't know how to install the DVD? See "DVD installation" on page xii.

Tools needed: ArcGIS 10.1 for Desktop.

1 Add shapefiles

1. Click the Add Data button to add agejoined. shp, which was created in chapter 5. This file is also in the C:\EsriPress\GIS20\12 folder.

2 Write a query

1. On the Selection menu, click Select by Attributes. →

2. Type the following query: "Percent" > 15. Click OK. Counties that meet the condition are highlighted with a bright blue line. →

3. Open the attribute table and notice the corresponding rows are also highlighted. (See "Export selected records" later in this chapter.) Close the attribute table.

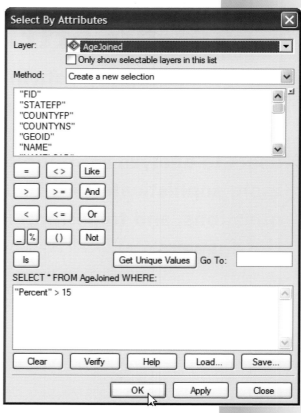

3 Create a new shapefile for selected counties

1. Right-click the layer name in the table of contents, click Data, and then click Export Data.

2. Navigate to your save folder. Name the file **seniors**. The Save as type should be Shapefile. This file will be used in the next exercise, so note where it is saved. Click Save and OK.

3. When prompted, select Yes to add the shapefile to the table of contents.

4. Clear the box next to agejoined in the table of contents so you are able to clearly see what is contained in the new file. This is how to create a "target area" map based on data.

EXPORT SELECTED RECORDS

It is easy to export a selection of records based on query results. After the query is completed, do the following:

1. In the table of contents, right-click agejoined and open the attribute table.

2. Click the Table Options tool arrow ⊟ ▾ on the Table toolbar, and then click Export. Save the file. The file type should be .dbf. This is a generic database file type that can be read by many software programs, including Excel and Access.

4 Erase the query

1. To erase the query, on the Selection menu, click Clear Selected Features. Alternatively, click the Clear Selected Features button ⊠. When the blue highlighting is gone, the query is erased and nothing is selected.

2. Close ArcMap. You do not need to save.

CHAPTER 13

Creating location queries

Location queries differ from attribute queries in that they do not involve selecting data. A location query is a geography query, not a data query. A location query involves selecting geographies within other geographies. It works equally well with point, line, or polygon data.

In this chapter, you will create a location query that selects cities within multiple selected counties.

Files and tools

Files needed: You will need seniors.shp (chapter 12) and tl_2011_01_place.shp (chapter 1). Or, if you prefer, you can install this book's DVD and access chapter files at C:\esriPress\GIS20\13. Don't know how to install the DVD? See "DVD installation" on page xii.

Tools needed: ArcGIS 10.1 for Desktop.

1 Add shapefiles

1. Click the Add Data button to add the Alabama place shapefile (tl_2011_01_place.shp) that was downloaded in chapter 1 (also available in the C:\EsriPress\GIS20\13 folder). Remember your file may have a different name and two-digit state FIPS code because this file is for Alabama.

2. Next, add seniors.shp, which was created in the last chapter.

2 Create a location query

The object here is to create a query that will select cities within the boundaries of the few counties in the Seniors shapefile.

1. On the Selection menu, click Select by Location.

2. For the section underneath Target Layer(s), select the box next to tl_2011_01_place.

3. For the source layer, select the Seniors shapefile.

4. For "Spatial selection method for target layer feature(s)" select "have their centroid in the source layer feature." Click OK. →

Notice only those places within senior counties are selected. All other cities outside of the senior counties are ignored.

5. Right-click the place layer name in the table of contents and open the attribute table. Here you are able to see which cities were selected.

    NOTE: For more on how to export just these selected records, see chapter 12.

6. Close the attribute table and ArcMap. You do not need to save anything.

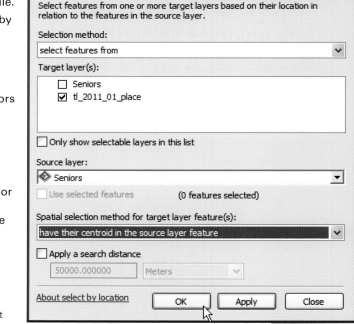

WHAT DOES "HAVE THEIR CENTROID IN" MEAN?

A centroid is the physical center point of geometry (in this case a polygon). By using the "have their centroid in" method, the physical center of the place must fall within the county boundary, not merely touch it or intersect it. This method is more conservative. You will never get polygons that are only slightly within the boundary because the center point of the polygon must fall within the target boundary.

Another popular tool to use for location queries is Intersect, which would include any place that intersects the county boundaries.

CHAPTER 14

Using geoprocessing tools

Geoprocessing refers to various GIS operations that manipulate spatial data. These tools include buffer, merge, union, append, and dissolve. The ability to perform very complex spatial calculations sets GIS apart from traditional cartography. It is the muscle of the GIS.

In this exercise, you will explore common geoprocessing tools by performing various tasks.

Files and tools

Files needed: All files for this exercise are accessible at C:\EsriPress\GIS20\14. You will need youthcenters.shp (C:\EsriPress\GIS20\14), tl_2011_48029_edges.shp (chapter 8), and states.shp (C:\EsriPress\GIS20\14). Don't know how to install the DVD? See "DVD installation" on page xii.

Tools needed: ArcGIS 10.1 for Desktop.

Buffer

A buffer is a map item that represents a uniform distance around a feature (point, line, or polygon). When creating a buffer, the user selects the feature to buffer around, as well as the distance of the buffer.

1 Add a shapefile

1. Open ArcMap. Click the Add Data button ✦ ▾ to add youthcenters.shp from the EsriPress folder, chapter 14. This file contains youth centers located in Bexar County, Texas. (It was created from the geocoded file from chapter 8 by performing an attribute query on the type column for youth centers.)

2 Buffer

1. To create a buffer of 1,000 feet around each of these youth centers, from the Geoprocessing list, select Buffer. →

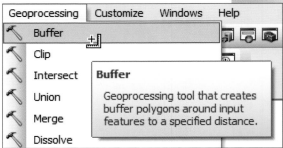

2. For Input Features, select youthcenters.

3. For Output Feature class, click the Browse button to navigate to your save folder. A new buffer file will be created. Name the file **buffers**, and then click Save.

4. For the linear input unit, type **1000**. The unit type should already be set to feet. Leave all other options as they are, and then click OK.

5. Drag the youthcenters layer into first position in the table of contents (make sure the List by Drawing Order button is clicked).

6. Zoom in closely to one of the buffers.

3 Add other files and change the symbology

1. Add street network tl_2011_48029_edges.shp, located in the EsriPress folder, chapter 14.

2. Turn labels on for streets, change the street color to gray, and make the buffer hollow with a thick red border (all items covered in previous chapters).

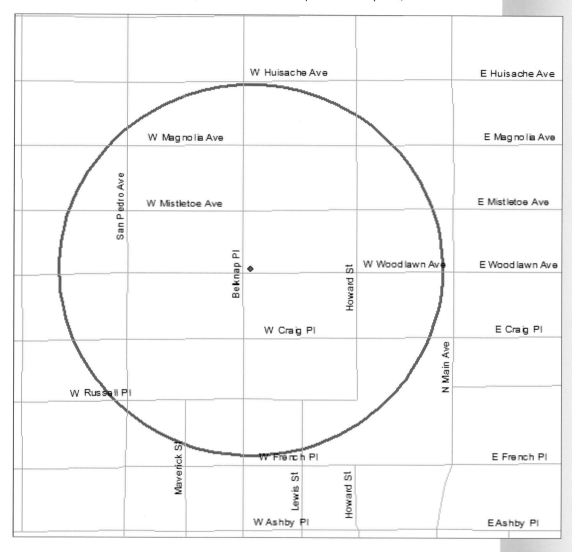

3. One other tool that may be useful here for double checking is the Measure tool 📐. Click this tool to enable it. Click the Choose Units button, and then click Distance. Feet should be checked.

4. Click the dot that represents the youth center and draw a line to the edge of the buffer. You should see a distance of 1,000 feet.

5. Close ArcMap. You do not need to save anything.

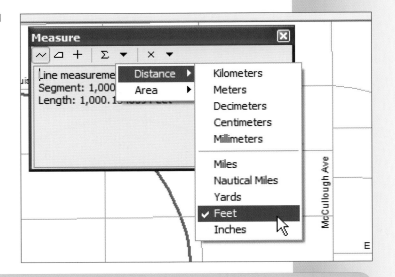

THE CHOOSE MEASUREMENT TYPE TOOL

Do you know spatial calculus? Me neither. That's OK because ArcMap does the work for us. The Choose Measurement Type tool gives us planar, geodesic, and a few other options. Planar uses 2D Cartesian mathematics to calculate distance and is used with projected coordinate systems, such as the one used in this chapter, state plane. It is designed to measure things on a flat surface. Geodesic measuring is a different type of measuring system that calculates the shortest distance between two points given the not perfectly round, ellipsoid shape of the earth.

The distance between two cities will not be the same if you (a) draw buffers using different underlying coordinate systems, and (b) measure using different measurement types. Geodesic is thought to be more accurate. In our example, there isn't much difference between planar and geodesic measurements because we created buffers for a small area and used a correct projection for the map.

Type **"Buffer Analysis"** into the Help menu and you will find a lot of information about this topic. One quick tip: the larger the area the more you should consider using a geographic coordinate system for buffering. This will minimize distance calculation errors.

Merging

It is important to start with a fresh mapping session here. If you did not close ArcMap from the buffer exercise, do so now. Alternatively, select File > New > Blank Map > OK. This will open a new session.

4 Merge

Example of Merge

1. Use the Add Data button to add the states shapefile from the EsriPress folder, chapter 14. This is a file of all the states within the United States. We also used this file in chapter 11 while editing.

2. Begin by making the shapefile editable. Click the Editor Toolbar button ⟍✎ to activate the Editor toolbar. Click the Editor button Editor ▾ and Start Editing.

3. Hold down the Shift key and, with the Edit tool, select multiple contiguous polygons such as three different states (in this example, Oregon, Washington, and Idaho). You can tell they are selected when all appear with bright blue shading.

4. Click Editor and Merge.

Because all three states will now take on the attributes of only one state, it is necessary to select which state that will be.

5. When you get the dialog box that says "Choose the feature with which other features will be merged," select one of the states. Oregon was selected for this example. Click OK. Notice that all three states have merged into one.

6. Open the attribute table. Notice that now neither Idaho nor Washington are present.

7. Close the attribute table. Click the Clear Selected Features tool 🔲 to clear the selection.

Union

The Union tool differs from the Merge tool in that each record in the attribute table is maintained for each merged item; however, the visible boundary between the polygons disappears. Visually a union looks like one big new polygon (same as a merge), but within the attribute table, a record is maintained for each polygon.

5 Union

1. The layer is still editable. Hold down the Shift key and, with the Edit tool ▶ , select another three contiguous polygons. In this example, Texas, New Mexico, and Oklahoma are selected.

2. Click Editor and select Union. When you get a dialog box that says "Choose a template to create feature(s) with," click OK.

3. Notice the three states now appear as one. However, they are still three distinct states. Click anywhere outside the states to get rid of the selection (blue highlight). Now click more or less in the area where Texas should be (or any one of the states you may have selected) and notice how the original outline of that one state appears selected. Visually it appears to be one boundary, but they are still three separate line items in the attribute table.

4. Open the attribute table to confirm you still have a line item for Texas, New Mexico, and Oklahoma. Close the attribute table. Click the Clear Selected Features tool to clear the selection.

5. Close ArcMap. You do not need to save anything.

Append

The Append tool combines multiple shapefiles into one shapefile.

It is important to start with a fresh mapping session here. If you did not close ArcMap from the union exercise, do so now. Alternatively, you can select File > New > Blank Map > OK. This will open a new session.

6 Append

1. Click the Add Data button to add PNW.shp and SW.shp from the EsriPress folder, chapter 14. The PNW shapefile contains three states in the Pacific Northwest. The SW file contains southwestern states.

The idea is to get these two separate files into one shapefile.

2. Click the ArcToolbox button 🗟 to open ArcToolbox. Expand the Data Management Tools toolbox, and then expand the General toolset. Double-click the Append tool.

3. Here we'll append the SW shapefile to the PNW shapefile. From the Input Datasets list, select SW. From the Target Dataset list, select PNW.shp. This is where SW will be appended. In the Schema Type field, leave the default TEST. Click OK. ➡

Input Datasets
◇ SW
Target Dataset
PNW
Schema Type (optional)
TEST

4. Turn off the SW shapefile. Notice all the states are now in one shapefile. The append was successful.

5. (Optional) In terms of file management, it could be confusing to call this file "PNW," when it includes not only the Pacific Northwest states but also the southwestern states. Here, you would likely create a new shapefile and give it a new name. Chapter 11 instructs you how to do this, but we will reiterate the process here. Right-click PNW, click Data, click Export Data, and navigate to your save folder. Rename the file and save it. The Save as type should be Shapefile.

6. Close ArcMap. You do not need to save anything.

APPENDING AND ATTRIBUTE TABLES

It is ideal if the attribute tables of all files in the append process are identical in terms of columns and column headings. If they are not, you are still able to append, but you must have at least one column in common. Search the Help menu for "Append Data Management" and read about Field Map Control.

Clip

Another tool you may find useful is Clip. You can clip one boundary by using the outline of another boundary. Let's try it.

7 Clip

1. Add states.shp and circle.shp from the EsriPress folder, chapter 14. Drag circle.shp into the first position in the table of contents, if it is not already there.

The idea here is we want to clip the states file to the boundary of the circle file.

2. On the Geoprocessing menu, click Clip.

3. For Input Features, select the file to be clipped (states.shp).

4. For Clip Features, select the file that will serve as the clipped boundary, in other words, the file that will be used to clip the first file (circle.shp).

Example of Clipping

5. For the Output Feature Class, navigate to your save folder. Name the new file **clipped**. Click save and OK.

6. The new file will automatically be added to ArcMap. Turn off all other shapefiles to clearly see the new clipped shapefile.

7. Right-click the new file and open the attribute table. Notice it takes on properties of the states file.

8 Right-click the circle layer and remove the file. Right-click the clipped layer and remove the file. Turn the states layer back on.

> **INTERSECT VERSUS CLIP**
>
> The Clip tool and the Intersect tool are similar. With the Clip tool, the new shapefile takes on the attribute properties of the original layer. It can only work with two layers at a time. Intersect is the same unless you have an Advanced license of ArcGIS for Desktop. With an Advanced license level, the resulting attribute table can assume the attribute properties of many layers.

Dissolve

Dissolve creates larger regions out of smaller regions.

8 Dissolve

1. The states shapefile should already be open in ArcMap.

2. Right-click the file and open the attribute table. Notice the Division column. These designations represent divisions of the United States as defined by the Census Bureau. Close the attribute table.

3. On the Geoprocessing menu, click Dissolve.

4. For Input Feature Class, select states. For Output Feature Class, navigate to your save folder. Name the new file **divisions** and click Save.

5. Select the DIVISION box in the Dissolve Field panel.

There is no data in this shapefile. If there were, you could select the fields under Statistics Fields and sum up all the values of the states in each division. For now leave it blank.

6. Click OK.

7. Close ArcMap. You do not need to save anything.

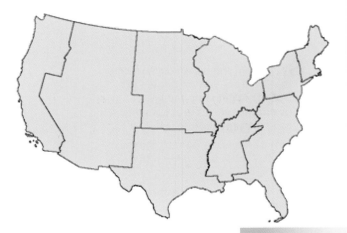

Example of Dissolve

CHAPTER 15

Creating geodatabases

You may encounter many types of files while working in ArcGIS. We have been working with shapefiles, the most basic and frequently used file type in ArcGIS. Shapefiles are used extensively not only in ArcGIS but with many other GIS software programs. Another commonly used file type is the geodatabase.

A geodatabase is like a container where you can store all files related to your GIS project. You can store multiple shapefiles, aerial photographs, spreadsheets, and many other types of files. To group these items together in one place is very convenient because it provides one point of access for all needed files. It is also handy if you would like to share these files with others. Putting everything in a geodatabase makes GIS data easier to manage and access.

If you are only using a few shapefiles for your GIS project, there is no need to convert it to a geodatabase. However, if your project begins to grow and use many files, it is worthwhile to convert the project to a geodatabase. In any case, it is good to be familiar with geodatabases, as many organizations distribute files in this file format. Even though you may never find the need to create one, it is likely that at some point you may encounter one.

In this exercise, you will explore how geodatabases work and work with ArcCatalog.

Files and tools

Files needed: tl_2011_48029_edges.shp (chapter 8), agejoined.shp (chapter 6), Agencies.xls (chapter 8), and Alamo.jpg (C:\EsriPress\GIS20\15). Don't know how to install the DVD? See "DVD installation" on page xii.

Tools needed: ArcGIS 10.1 for Desktop.

1 Open ArcCatalog

ArcCatalog is the library system of ArcMap. You can browse, copy, delete, and organize files in ArcCatalog. ArcCatalog is also where you create file and personal geodatabases.

1. On the Windows menu, click Catalog. →

   You can also click the Catalog button 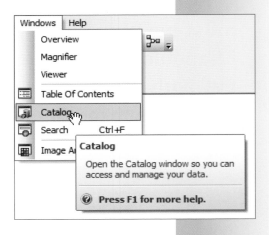.

   Notice a new window has opened on the right side of the screen. This is ArcCatalog. ArcCatalog looks slightly different in older versions of the software.

2 Explore ArcCatalog (optional)

If you are not already familiar with ArcCatalog, take this opportunity to explore it.

1. Expand the Folder Connections folder. Notice that saved connections to folders are located here. Right-click Folder Connections and notice you are able to connect to folder using this method, instead of clicking the Connect to Folder button.

2. Assuming you have a file available under Folder Connections, right-click the file name and notice the menu of options here. You can create new folders as well as copy, delete, rename, and export files. This is very handy for managing files.

3 Create a file geodatabase

The basic process for creating a geodatabase is to first create the geodatabase, and then import or export files to the geodatabase.

1. In ArcCatalog, under Folder Connections, navigate to your save folder. Right-click the save folder, click New, and then click File Geodatabase.

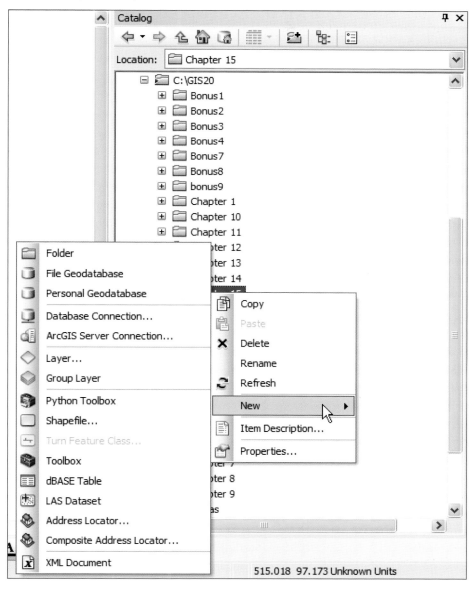

2. When prompted, rename the file **gdb.gdb**. When you are finished, click in the white space to deactivate the text tool.

4 Import shapefiles to the geodatabase

1. Right-click gdb.gdb, click Import, and click Feature Class (multiple). This will import shapefiles to the geodatabase.

2. In the Import Features field, navigate to the chapter 15 folder in the EsriPress folder and select tl_2011_48029_edges.shp, which is the Bexar County street network. Then, navigate to chapter 15 and import agejoined.shp. Click OK. It may take a moment.

3. Once the operation is complete, expand the geodatabase in ArcCatalog and notice these two shapefiles have been added to the geodatabase. ➜

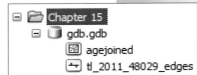

5 Export aerial photos to the geodatabase

The import function does not work with aerial photos, nor does it work with Excel spreadsheets. Instead we must first navigate to those files, and then export them to the geodatabase.

1. In ArcCatalog, navigate to the georeferenced Alamo aerial photo (Alamo.jpg) located in C:\EsriPress\GIS20\15.

2. Right-click the file, click Export, and then select "Raster to Different Format."

3. The Input Features field should already contain the file to export to the geodatabase.

4. In the Output Location field, navigate to gdb.gdb. Double-click the file and type a new name. Call it **Alamo**. (If there are spaces in the file name, an error will be generated and the operation will fail.) The Save as type should be Raster Datasets. Click Save.

5. Leave all of the other options and click OK. It will take a minute to process.

6. Once it finishes, double-click gdb.gdb in ArcCatalog and notice the file has been added to the geodatabase. It will also open in ArcMap.

> ### INCREDIBLY USEFUL TIP
> You can also drag files from ArcCatalog into the table of contents to easily open files.

6 Export an Excel table to the geodatabase

1. In ArcCatalog, navigate to C:\EsriPress\GIS20\15 and double-click Agencies.xls.

2. Right-click the worksheet Organizations$, click Export, and then select To Geodatabase (single).

3. The Input Features field should already contain the file to export to the geodatabase.

4. In the Output Location field, navigate to gdb.gdb. Do not give it a new name here. Click Add.

5. In the Output Table field, type a new name. Call it **Addresses** (no spaces in file name). Click OK. It will take a minute to process.

6. Once it finishes, double-click gdb.gdb in ArcCatalog and notice the file has been added to the geodatabase. It will also open in ArcMap, but you will not be able to see it because it doesn't have any geographic attributes.

7. Close ArcMap. You do not need to save anything.

GEODATABASES SERVED THREE WAYS

ArcGIS supports three types of geodatabases: file (.gdb), personal (.mdb), and ArcSDE (.sde).

For one person, or small workgroup projects, the file geodatabase is the best choice. It has no size limitations. The personal geodatabase is still used, but this is an older file type and has a 2 GB size limitation.

The ArcSDE geodatabase (SDE stands for Spatial Database Engine) is for large workgroups. It is not accessible with a Basic license. This type of geodatabase would be created and administered by a GIS manager. SDE has some big advantages. Files can be locked so they cannot be edited by different people, helping to maintain data integrity. Also, SDE can store and draw raster data, such as aerial photographs, very efficiently. Another key advantage is that multiple people can work with, and even edit, data simultaneously, unlike in file and personal geodatabases.

CHAPTER 16

Joining boundaries

In chapter 5 we joined a data table to a shapefile. In this chapter, we will perform a spatial join, which involves combining two shapefiles (and their data tables) into one. For example, what if you wanted to figure out which census tracts are in which neighborhoods in your city? One way to accomplish this is by clicking each tract on a map and writing down the neighborhood it is in. But this would take much too long. Instead, you can perform a quick spatial join to combine tract and neighborhood shapefiles. The end result will be an attribute table that includes both tract and neighborhood data. This is an extremely powerful tool that is useful in many situations.

In this exercise, you will perform a spatial join between Alabama city and county shapefiles resulting in an attribute table that identifies the county in which each city is located.

Files and tools

Files needed: You will need alabamacounties.shp (chapter 1) and tl_2011_01_place. shp (chapter 1). Or, if you prefer, you can install this book's DVD and access chapter files at C:\EsriPress\GIS20\16. Don't know how to install the DVD? See "DVD installation" on page xii.

Tools needed: ArcGIS 10.1 for Desktop.

1 Add shapefiles

1. Open ArcMap. Click the Add Data button ✛ ▾ to add alabamacounties.shp (your shapefile will have your state's name) and tl_2011_01_place.shp from chapter 1. These files are also accessible via C:\EsriPress\GIS20\16.

2 Activate the Spatial Join tool

1. Click the ArcToolbox button 🗔 to open ArcToolbox.
2. Expand the Analysis Tools toolbox, and then expand the Overlay toolset.
3. Double-click the Spatial Join tool.

3 Fill in spatial join options

1. For Target Features, select the shapefile to which you want the data appended. This will be the places shapefile tl_2011_01_place.shp.
2. For Join Features, select the file you would like to append. This will be the county shapefile alabamacounties.shp.
3. For Output Feature Class, navigate to your save folder. Name the file **spatialjoin**. Click Save.
4. In the Join Operations field, select Join_One_to_One.
5. In the Field Map section, you will need to indicate that you would like each county's name to appear in the results column separated by a comma. Right-click Name_1 (Text) and click Properties. This is important because it is possible for a city to span over multiple counties. You would like the results to indicate each county that the city falls within, even if it falls within multiple counties.

6. Change the length from 100 to 254, the maximum characters allowed. Select Join for Merge Rule and type a comma for the Delimiter field. If you do not select Join for the merge rule type, then only the first encountered county would be the listed county, regardless of whether it spanned multiple counties. Click OK. →

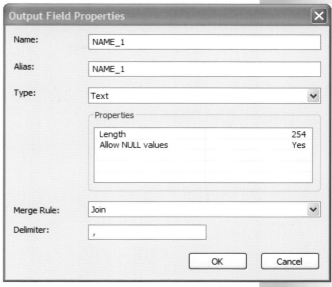

7. For the Match Option, select Intersect. There are other options here, but you would like to know which county or counties each city intersects. Click OK. The operation may take a minute. The new spatialjoin shapefile will be added to the table of contents.

4 Open the attribute table and view results

1. In the table of contents, right-click the new shapefile name and open the attribute table.

2. Notice the Name_1 column now indicates which county or counties the city falls in. For example, Leeds city falls within St. Clair, Jefferson, and Shelby counties. The city is listed in the Name column toward the beginning of the table. It is necessary to scroll right to see the Name_1 column and see the county information. →

3. Close the attribute table and ArcMap.

Franklin
Franklin
St. Clair,Jefferson
St. Clair,Jefferson,Shelby
St. Clair,Jefferson
St. Clair

CHAPTER 17

Working with aerial photography

Aerial photos are great for displaying a photographic snapshot of what is happening on the ground in any particular area. They are not considered "smart maps" because you cannot click the map and get information about a particular feature. Aerial photography, a type of raster data, can be particularly useful when combined with shapefiles or other vector data. For example, utility companies may want to view a picture of a five-block area, and then overlay (in shapefile format) utility poles (points perhaps collected from a GPS) and associated information about those poles such as electricity generated or malfunctioning power lines. This type of information changes constantly and can easily be updated in a shapefile, but not in an aerial photograph. Therefore, learning to work with both types of files, especially in combination, is useful.

Files and tools

In this exercise you will focus on aerial photographs; however, these techniques would also work well with a scanned map. You will import an aerial image of the area around the Alamo, located in downtown San Antonio, Texas, and overlay a street network. You will learn how raster and vector data work together by georeferencing raster data.

Files needed: All files for this chapter are accessible at C:\EsriPress\GIS20\17. You will need Alamo.jpg (C:\EsriPress\GIS20\17) and tl_2011_48029_edges.shp (chapter 8). Don't know how to install the DVD? See "DVD installation" on page xii.

Tools needed: ArcGIS 10.1 for Desktop, an Internet connection.

1 Add the shapefile

1. Add tl_2011_48029_edges.shp by clicking the Add Data button ✛ ▾ and navigating to C:\EsriPress\GIS20\17. This shapefile contains streets in Bexar county, Texas, where the Alamo is located.

> ### INCREDIBLY USEFUL TIP
> For this type of project, always add the shapefile before adding the aerial photo.

2 Add an image to your map

1. Click Add Data button again and navigate to the aerial photograph located in C:\EsriPress\GIS20\17.

2. When the dialog box appears that warns of an "Unknown Spatial Reference," click OK. The image will be added to the table of contents, but will not be visible in data view yet. Three bands will be added under the photograph.

> ### NO SPATIAL REFERENCE: WHY THIS IS A PROBLEM
> In this exercise, we want to display the streets file on top of the image file. Right now the image file has no spatial reference, so the program does not know where to put it in relation to the streets file.

3 Move between the two layers

1. Even though you have added the image, it will not be immediately apparent. To view it, in the table of contents, right-click the aerial photograph's layer name and click Zoom to Layer.

2. To view the streets, do the same thing but with the streets layer.

4 Identify four intersections in the aerial photograph

You must identify at least three points (the more the better, though) on the image that will serve as anchor points for spatial referencing. To make your task easier, four intersections have been identified and labeled in the illustration.

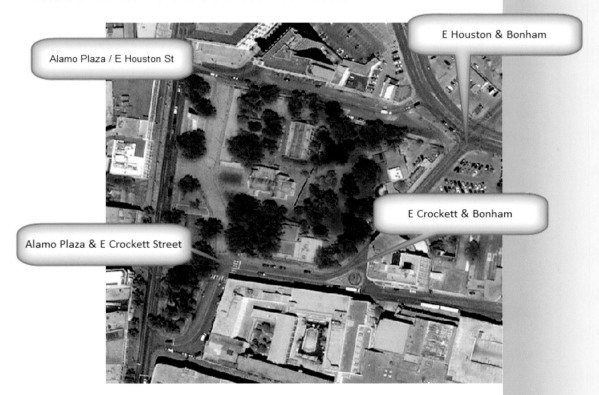

The four intersections identified are the following:

1. Alamo Plaza & E Houston Street (northwest corner)
2. Alamo Plaza & E Crockett Street (southwest corner)
3. E Crockett & Bonham (southeast corner)
4. E Houston & Bonham (northeast corner)

5 Identify the first intersection on the street network in ArcMap

1. In the table of contents, right-click the street network layer and click Zoom to Layer. It is important to make sure you are looking at the street network before proceeding.

2. Click the Find tool 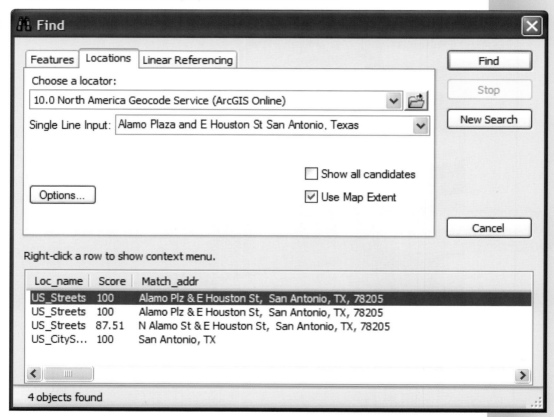, which looks like a pair of binoculars, next to the Identify tool. Click the Locations tab.

3. Click the arrow next to "Choose a locator" and select the first option, "10.0 North America Geocode Service (ArcGIS Online)." In the Single Line Input field type **Alamo Plaza and E Houston St San Antonio, Texas**. Select the Use Map Extent box. Click Find. The results will display in the box at the bottom.

NOTE: This shapefile has been previously projected using StatePlane Texas South Central FIPS 4204 Feet. For more on projections, see chapter 3.

4. Click the first intersection in the Address Description section. Notice the spot on the map flashes, but the view is still too far out to see much.

5. Right-click the first intersection and click Add Point. This will place a point at that intersection, so it is very easy to see.

> ### INCREDIBLY USEFUL TIP
> *If Add Point is unavailable, make sure you are in data view. If you are not, go to View > Data View to switch from layout view to data view.*

6. Close the Find tool. Use the Zoom In tool to see the placement of the dot.

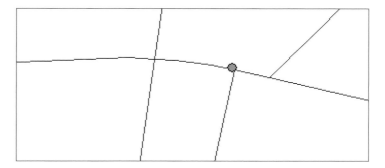

6 Fit to display

1. On the Customize menu, click Toolbars, and then click Georeferencing.

2. Click the Georeferencing button Georeferencing ▾ and select Fit to Display. This will display both the aerial photo and streets in the same view, but they are not aligned.

3. Right-click the streets layer name in the table of contents, click Properties, and then click the Symbology tab. Click the symbol swatch and change the streets to a color and width that are easier to see.

7 Add control points

Adding control points allows you to align the image with the shapefile layer.

1. Turn the streets layer off and figure out where the intersection of Alamo Plaza and E Houston Street is on the aerial photo. (Hint: It is in between two white cars in the northwest corner of the map.) Turn the street layer back on.

2. On the Georeferencing toolbar, click the Add Control Points tool ⤡.

3. With the Add Control Points tool activated, click the place where the intersection is on the image first. A line will appear. Drag that line to the intersection on the street network. It must be done in this order. Click the image first, then the shapefile. The intersections should line up somewhat. This is control point 1. (If you haven't already, you can delete the dot you placed earlier by selecting the default pointer. Click the dot and click Delete on your keyboard.)

WHAT HAPPENS IF I MESS UP?

The View Link table catalogs all control points. If you accidentally click the wrong point, no problem! Click the Delete Link button ⊹⊼, select the point, and delete it. Alternatively, on the Georeferencing toolbar, you can click the View Link Table button ⊞. Click the point you would like to delete in the table, and click Delete on your keyboard. This will delete the point. You can then close the table or keep it open.

4. Add another control point for the intersection in the southwest corner, Alamo Plaza and E Crockett Street. Remember, click the image first, then the shapefile. The intersections should line up. This is control point 2.

5. Do the same thing for the other two intersections. Throughout this process you will likely need to zoom in and out to see more clearly. The mouse scroll wheel is great for this.

8 View Link table

1. After all four points are placed, on the Georeferencing toolbar, click the View Link Table button ⊞. Scroll right and notice the residual column at the end. This indicates how closely the two files are aligned. A total score is given in the upper right corner in the Total RMS Error field. RMS means root mean squared. The closer this number is to zero the better. No good guidelines on what the RMS should be exist. In a perfect world the RMS would be zero. Does that happen in real life? Never. Do the best you can. Close the table. ➡

Total RMS Error:	Forward: 19.51

9 Update georeferencing

1. Once you have finished adding points and working with the data, on the Georeferencing toolbar, click the Georeferencing button.

2. Click Update Georeferencing. Now the image can be used in conjunction with other shapefiles for this area.

I HAVE A SCANNED MAP. HOW CAN I TURN THAT INTO A SHAPEFILE?

It is impossible to turn a map into a shapefile. That would be magic. You can follow the steps in this exercise, except substitute a scanned map for the aerial photograph. That would let you digitize it and overlay other layers on top of it. You could also use editing tools to create a new shapefile and sketch a boundary of something like a target area. For more on how to do that, see chapter 11.

CHAPTER 18

Creating reports

If you use GIS to provide information to other people, you may find creating reports a useful feature. Reports can give your reader a lot of information and provide credibility for the map's data. ArcGIS for Desktop has a new and improved report builder. Report data can either be included as a part of your map's layout (if it is a small amount of data) or attached as a technical addendum.

In this exercise, you will create and export a basic data report.

Files and tools

Files needed: You will need agejoined.shp (chapter 5). Or, if you prefer, you can install this book's DVD and access chapter files at C:\EsriPress\GIS20\18. Don't know how to install the DVD? See "DVD installation" on page xii.

Tools needed: ArcGIS 10.1 for Desktop.

1 Add shapefiles

1. Click the Add Data button ✛ ▾ to add agejoined.shp, created in chapter 5, to your map. This file is also in C:\EsriPress\GIS20\18.

2. A new option in ArcGIS 10.1 for Desktop allows you to create a report for only what's visible in the map frame. To use this new feature, zoom in fairly close to your map, displaying around ten counties.

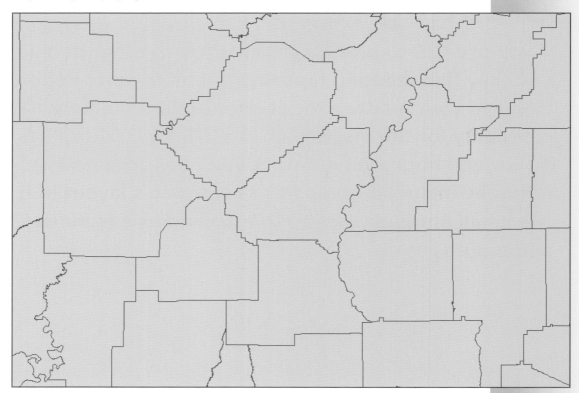

2 Open the attribute table

1. To refresh your memory about the types of data we have in this shapefile, in the table of contents, right-click agejoined and click Open Attribute Table.

2. Scroll far to the right and notice that each county has an entry for total population, total number of seniors, and the percentage of seniors in the population. These columns will be the basis for our report.

3 Create a report

1. Click the Table Options button, click Reports, and then click Create Report.

Table

		YFP	COUNTYNS	GEOID	NAME	NAMELSAD
🔍	Find and Replace...		00161528	01005	Barbour	Barbour County
🔲	Select By Attributes...		00161583	01113	Russell	Russell County
	Clear Selection		00161569	01087	Macon	Macon County
🔲	Switch Selection		00161575	01097	Mobile	Mobile County
🔲	Select All		00161541	01031	Coffee	Coffee County
	Add Field...		00161555	01059	Franklin	Franklin County
			00164997	01115	St. Clair	St. Clair County
🔲	Turn All Fields On		00161576	01099	Monroe	Monroe County
✓	Show Field Aliases		00161533	01015	Calhoun	Calhoun County
	Arrange Tables ▶		00161566	01081	Lee	Lee County
			00161560	01069	Houston	Houston County
	Restore Default Column Widths		00161589	01127	Walker	Walker County
	Restore Default Field Order		00161543	01035	Conecuh	Conecuh County
			00161556	01061	Geneva	Geneva County
	Joins and Relates ▶		00161592	01133	Winston	Winston County
	Related Tables ▶		00161530	01009	Blount	Blount County
			00161567	01083	Limestone	Limestone County
📊	Create Graph...		00161579	01105	Perry	Perry County
	Add Table to Layout		00161531	01011	Bullock	Bullock County
			00161526	01001	Autauga	Autauga County
🔄	Reload Cache		00161562	01073	Jefferson	Jefferson County
			00161590	01129	Washington	Washington County
🖶	Print...				Escambia	Escambia County
	Reports ▶	📋 Create Report...		Henry	Henry County
	Export...	📇 Load Report...		Clarke	Clarke County
	Appearance...	▶️ Run Report...			

Create a report, load an existing report from a file, or regenerate a report so it reflects the latest data.

	27	Polygon	01	093	00161573	010
	28	Polygon	01	055	00161553	010

2. Under Available Fields, scroll down and select County. Click the right arrow to deposit it into the Report Fields section. Do the same thing with the Percent column. The report will have two columns.

3. Click the Dataset Options button, select Visible Extent, and then click OK. This will create a report for only those counties displayed in the map frame. Click Next twice.

4. When you get to the field sorting options, click the arrow under Fields and select the Percent column. Click the arrow under Sort and select Descending. This will put higher percentages at the top of the column.

5. Click the Summary Options button. Summary Options calculates the average, which often is very helpful. It also does a few other calculations.

6. Select the Avg box, click OK, and then click Next.

7. The next step relates to the report layout. Leave the defaults and click Next.

8. The next step relates to styles. Review the styles. Good options for styles are New York and Simple. Select New York and click Next.

9. Type **Percentage of Seniors** as the title of the report, and then click Finish. Your data will display in report format.

4 Export report

You may want to export your report and attach it to your map, or you may want to export it to Excel for further analysis.

1. On the toolbar at the top of the report, click the Export Report to File button.

2. Click the Export Format arrow and notice several options. Select Portable Document Format (PDF).

3. In the File Name field, click the ellipsis and navigate to where you would like to save the file. Name it **report**. Click Save and OK. You have saved your report in PDF format.

4. Close the report and close ArcMap. You do not need to save anything.

ADD A REPORT TO ARCMAP

You can also add a report directly to a layout in ArcMap by clicking the Add Report to ArcMap Layout button.

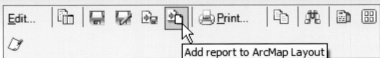

ATTRIBUTE QUERIES IN REPORTS

The ability to write an attribute query and have the results of that query make up the report is a useful report feature available in ArcGIS. When you select the columns to include, if you click Dataset Options, you can select only certain records to include in the report. You also have the option to select Definition Query as the option and a query wizard will open. Here you can write an attribute query that will select records based on your own criteria. See chapter 12 for more on writing attribute queries.

CHAPTER 19

Sharing work

It is easier than ever to share maps and all the pieces
that make up those maps. You can share an entire
workspace or just a few layers. With new package
functionality available in ArcGIS, it is easy to send
layers and projects via e-mail. It is also easy to upload
data to ArcGIS Online, which is covered in chapter 20.

In this exercise, you will learn how to package maps, layers, and tools for the purpose of sharing your work with others. This chapter focuses on offline saving and sharing options. Chapter 20 shows you how to share your work using ArcGIS Online.

Files and tools

Files needed: You will need seniors.mxd (chapter 6), states.shp (C:\EsriPress\GIS20\19), and circle.shp (chapter 14). Don't know how to install the DVD? See "DVD installation" on page xii.

Tools needed: ArcGIS 10.1 for Desktop.

1 Open the project

1. Open ArcMap. On the File menu, click Open and open seniors.mxd, which was created in chapter 6. Files for this exercise are in C:\EsriPress\GIS20\19.

Map packages

2 Save the map package to your desktop

This method saves the map package (.mpk) to your desktop. You can then e-mail the map package or save it to a flash drive. It contains all files for this project. The end user will see exactly what you see in ArcMap.

1. On the File menu, click File > Share as > Map Package.

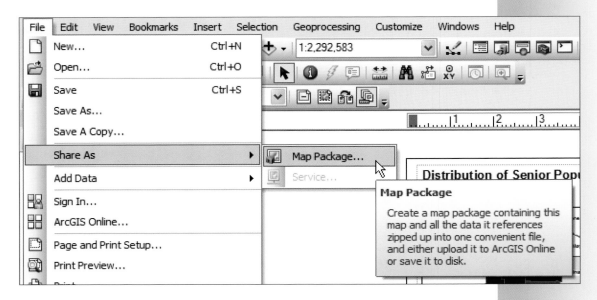

First we must input some information about this map. You are not able to share map packages without this information.

2. Select Item Description on the left and fill in the required fields. For the summary, type **Alabama senior population by county 2010** (with your own state's name). For credits, type **2010 US Census**. For tags, type **seniors Alabama**. Select the box to update missing metadata based on this information. →

Item Description

Summary (required):

Alabama senior population by county 2010

Tags (required):

seniors alabama

Choose Your Tags...

Description:

Access and Use Constraints:

Credits:

2010 US Census

☑ Update missing metadata in document based on item description.

3. Click Map Package on the left. Select Save Package to File, navigate to your save folder, leave the name as seniors.mpk, and click Save.

4. In the upper right corner, click the Share button. Click OK when the operation succeeds. A new map package file is created on your desktop. This contains all the files that make up this project.

5. Close ArcMap. Navigate to the senior.mpk file in your save folder. Double-click the file. It will open in ArcMap.

Layer packages

Layer packages (.lpk) allow you to save individual layers, not the entire workspace. This will save the layer (including color shading and labeling) along with the underlying shapefile, making the layer package a very handy option.

3 Save a layer package to your desktop

1. Right-click the agejoined layer name in the table of contents and click Properties.

2. Click the General tab. In the Description panel, type **Alabama Seniors** (there must be a description here). Click OK.

3. Right-click the agejoined layer name in the table of contents and click Create Layer Package.

4. Select Item Description on the left and fill in the required fields. For the summary, type **Alabama senior population by county 2010** (with your own state's name). For tags, type **seniors Alabama**.

5. Click Layer Package on the left. Select Save Package to File, navigate to your save folder, leave the name as agejoined.lpk, and click Save.

6. In upper right corner click the Share button. Click OK once the operation succeeds. A new layer package file is created on your desktop. This contains this layer plus its data.

7. Close ArcMap. You do not need to save anything. Navigate to the agejoined.lpk file in your save folder. Double-click the file. It will open in ArcMap. It may take a moment.

Geoprocessing packages

Geoprocessing packages (.gpk) allow you to share your workflow with others. Chapter 14 covered some basics of geoprocessing. Imagine being able to zip that whole session and send it to someone else. That's what a geoprocessing package does. All resources (models, scripts, data, layers, and files) needed to reexecute the tools are included in the package.

4 Save geoprocessing package to your desktop

1. Let's use the Clip to create a geoprocessing package. Open states.shp and circle.shp again from chapter 14, also in C:\EsriPress\GIS20\19. Drag circle.shp into the first position in the table of contents.

You want to clip the states file to the boundary of the circle file.

2. Under the Geoprocessing menu, select Clip.

3. For Input Features, select the file to be clipped (states.shp).

4. For Clip Features, select the file that will serve as the clipped boundary, meaning the file that will be used to clip the first file (circle.shp).

5. For the Output Feature Class, navigate to your save folder, name the new file **clippedagain**, click Save, and then click OK. The new file will automatically be added to ArcMap.

6. On the Geoprocessing menu, click Results.

7. In the Results window, right-click Clip, click Share as, and then click Geoprocessing package.

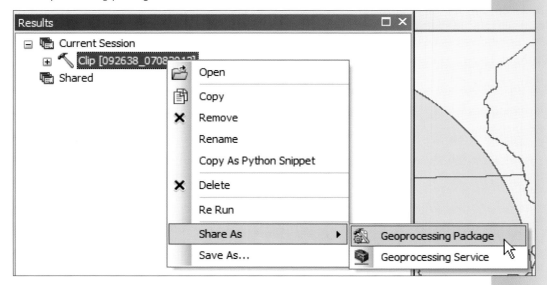

> ### INCREDIBLY USEFUL TIP
> *Here you are also able to publish as a service if your organization maintains an ArcMap Server.*

8. Select Save Package to File. Click the Browse button and navigate to your save folder, leave the name as clip.gpk, and click Save.

9. Clear the Include Enterprise Geodatabase box. Click Share. Click OK when the results are successful.

5 Rerun the geoprocessing package

1. Rerun the geoprocessing package. On the File menu, click New, Blank Map, and OK. This will open a clean workspace.

2. Open ArcCatalog 🗐, navigate to your save folder, and notice a new file called clip.gpk.

3. Drag clip.gpk into the empty mapping window. Click the plus sign next to the Clip layer in the table of contents to expand the layer. The three pertinent shapefiles are now available in the table of contents. Close ArcCatalog.

4. In order to rerun these results, remove the clipped file so we can re-create it by rerunning the geoprocessing. Right-click the clippedagain shapefile and remove it.

5. On the Geoprocessing menu, click Results. The Results window opens. Click the plus sign next to Clip to expand it. Right-click the second Clip link and click Re Run. Give the tool a moment to work. After it is finished, the newly clipped shapefile will be added to the table of contents. ➔

6. Expand all the menus in the Results window and notice all the information available about the operation. Close the Results window.

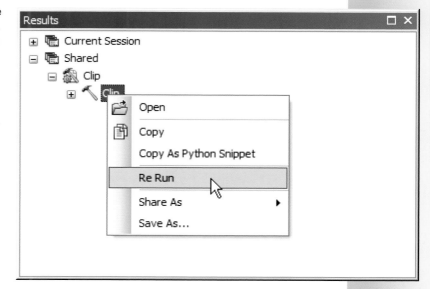

> ### PACKAGE TOOLBAR
> In ArcToolbox under Data Management Tools, there is a Package tool. This tool provides the options discussed here as well as a few others.

LOCATOR PACKAGES

Another package tool provides the ability to create address locator packages (.apk) and share them with others so the package may be used during their geocoding sessions. A lot can go into creating an address locator, especially if it is a composite address locator with a comprehensive alias table. This tool is in ArcToolbox > Data Management Tools > Package > Package Locator.

Publishing maps

You have several good options to publish and share maps. ArcGIS Online provides a space for users to upload and download maps, data, packages, and all sorts of other useful stuff. Think of it as a community for GIS users.

Geoenabled PDFs are another useful way to publish (in print form) your maps. The "geoenabled" part allows you to package the underlying data with maps, which in turn, enables you to click the map and access attribute data.

In this exercise, you will explore ArcGIS Online and make a geoenabled PDF file.

Files and tools

Files needed: You will need agejoined.shp (chapter 5), seniors.mxd (chapter 6), and seniors.mpk (chapter 19). Or, if you prefer, you can install this book's DVD and access chapter files at C:\EsriPress\GIS20\20. Don't know how to install the DVD? See "DVD installation" on page xii.

Tools needed: ArcGIS 10.1 for Desktop, an Internet connection, and a browser.

1 Create an ArcGIS Online personal account

The ArcGIS Online personal account is a free account and is required to access many features in ArcGIS Online.

1. Open an Internet browser and navigate to **http://www.arcgis.com**.

2. Click the Sign In link, and then click the Create Personal Account button.

3. Fill in the required fields. You must review and accept the terms of use to create an account.

4. Feel free to write a short description about yourself, make your profile public or private, and set language and location preferences. You can also skip this step.

ArcGIS Online

2 Explore the ArcGIS Online Gallery

1. Click the Gallery Link.

2. In the search field in the upper right corner type **USA Median Household Income** and click the Search button. Click the first link, USA Median Household Income.

3. Click the Open button and select the first link, Open in ArcGIS.com map viewer. →

3 Find a location and geographic scale

1. In the upper right corner in the search field, type your address or city name. For this exercise Portland, Oregon, is used.

2. On the left, in the table of contents, click the Show Map Legend button. →

3. Practice zooming in and out by using your mouse's scroll wheel (or the Zoom In and Zoom Out tools). Notice how the geography changes from block group, to tract, to county, and to state.

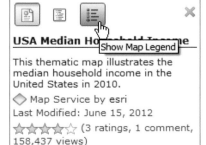

4 Access data

1. Click any part of the map. Notice an information box appears and the geography is highlighted in blue. →

5 Save your map

1. Click Save > Save As.
2. Fill in the options and click Save Map.

6 Share using social media

1. Click the Share button (top, center).
2. Select the box to allow everyone to view your map.
3. If you have a Facebook or Twitter account, you can post your map directly from here.

7 Share by web application

1. Click the Make Web Application button. This feature allows you to make and store maps on Esri's servers. You can store up to 2 gigabytes for free, after that, you must purchase a plan.

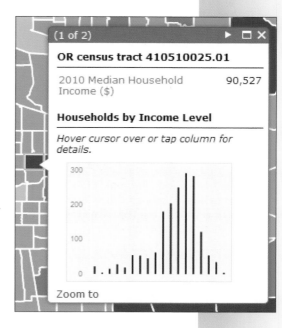

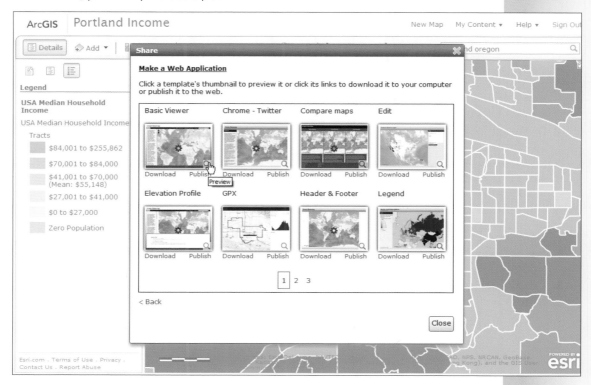

2. Under the Basic Viewer, click Publish, and then click Preview. Use this tool to see a preview of what your map will look like.

3. Notice all the great options here, including the ability to move the map, zoom, type in an address, as well as many other features. Once you are finished, close the template.

Now let's publish the map.

4. Click the Publish link under the Basic Viewer template.

5. Fill in the options and select Save & Publish. Click the Close button on the next box.

6. Notice a new browser with your map. Click the Share button and select e-mail. This works with Microsoft Office. You can also copy and paste the link at the top in order to share with others. ➡

EMBED YOUR MAP IN A WEBSITE

On the Share menu, you have the option to embed in a website. This provides the code to create an Iframe. To make it work, you must copy the code into a blank HTML page that you can then upload to a server and publish on your website. That's all you have to do in order to get a map onto your website.

8 Upload to ArcGIS Online

1. On the ArcGIS Online site, click the My Content link, and then select ArcGIS.com Home. From the main interface, click the My Content button. ➡

2. On the left, click New to create a new folder. Call the new folder **Desktop** and click Create.

3. Notice on the right, you can delete, share, and move maps. You can also add new maps and packages here from your desktop. Let's do that. First click the Desktop folder on the left, and then click Add Item.

4. In the first field leave the default, On My Computer. For the File section, navigate to your save folder and select seniors.mpk (also in C:\EsriPress\GIS20\20). For title, type **Seniors - <your name>** (example: Seniors - Gina Clemmer). For tags, type **seniors, <your name>**. Tags are what will enable you (and others) to find your file in ArcGIS Online. Click Add Item.

5. Click the Seniors - <your name> link under Title. ➡

6. Click the Share button, select the box to share it with everyone, and click OK. →

7. Click ArcGIS in the upper left corner to go to the homepage.

8. At the top of the page, click Show and select All Content. →

9. In the search field, type your name and click the Search button. Your file should display as a result.

ARCGIS ONLINE: FREE OR FEE?

ArcGIS Online is free to use. You can store up to 2 GB for free with just a personal ArcGIS Online account. After that point, an organization subscription is necessary. For more information, go to http://www.esri.com > Products > ArcGIS Online.

9 Download from ArcGIS Online

1. Click the title link.

2. Click the Open button and select download. You may not want to save it in the same save folder because the seniors.mpk file already exists there. If you do, it is fine to overwrite that file. After it downloads, double-click the map package file to open it in ArcMap.

INCREDIBLY USEFUL TIP

Different file types provide slightly different options here. If the option to Open in ArcMap is available, it will open automatically in ArcMap. You can use this option and skip the saving to desktop option.

3. Close ArcMap. You do not need to save anything.

BASEMAPS

Basemaps can be accessed in ArcMap by clicking the Add Data button.

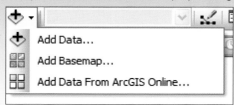

Basemaps do not contain attribute information and are not shapefiles. They are visual images of maps, but you cannot join to them or retrieve attribute data from them. They are helpful to give contextual information to your map, but not very helpful for analysis.

GROUPS: PUBLIC AND PRIVATE

ArcGIS Online provides the valuable resource of allowing users to create groups that can be public or private. For private groups, members can join by invitation only and no search will identify the group. This is great for internal workgroups. Public groups on the other hand will turn up in search results and anyone can apply to join the group. You can search for groups to join by clicking the Group button and searching.

Publishing geoenabled PDF maps

Another good way to share maps is to export them as a geoenabled PDF. In Adobe Acrobat Reader 6.0 and higher, you can turn individual map layers on and off. With Adobe Reader 9.0 you can use many new tools that work with ArcMap. The steps below work with either version, but it is a good idea to upgrade if you haven't. Adobe Acrobat Reader is free software.

10 Open the project

1. Open ArcMap. On the File menu, click Open. Navigate to seniors.mxd in your save folder. It is also in C:\EsriPress\GIS20\20.

11 Modify the attribute table

Select only a few columns to bring over with the PDF since we don't need all columns. Here's how:

1. In the table of contents, right-click the layer name and click Properties.
2. Click the Fields tab and notice all columns have a check mark next to them. If you were to leave it like this, all these columns of data would be exported with the PDF.
3. Click the Turn All Fields Off button ⬚.
4. Select the County, Pop, Seniors, and Percent boxes. These are the only columns that will be included. Click OK.

5. In the table of contents, right-click the layer and click Open Attribute Table. Notice that now you only see the selected columns.

6. Close the attribute table

12 Export the map as a geoenabled PDF

1. On the File menu, click Export Map.

2. Set the save location and select Save as type as PDF. Name the file **seniors.pdf**.

3. Click the Advanced tab. From the Layers and Attributes list, select Export PDF Layers and Feature Attributes. This will enable the underlying attribute data to be exported with the map. →

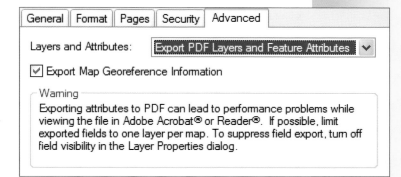

4. Click the Security tab. This is a new feature in ArcGIS 10.1 for Desktop. Here you are able to password-protect the map if you choose. Click Save.

13 Look at layers in Adobe Reader

1. Open the file in Adobe Reader, the free PDF viewing software. Adobe Reader 10.1.3 is used in this exercise, but any version 6.0 or higher will work.

2. In Adobe Reader, on the navigation menu on the left, click the Layers button. It looks like sheets of paper stacked on top of each other.

3. Click the plus sign next to Layers to expand the menu. Click the second plus sign as well.

4. Although we only have one layer open here, click the button that looks like an eye 👁 to turn map elements off and on. If you had multiple layers, you could turn specific ones off and on to suit your needs.

14 Access attribute information from within the PDF

1. On the Edit menu, click Analysis, and then click the Object Data tool.

2. On the map, double-click any county. It will become outlined in red.

Attribute information for the selected county displays on the left at the bottom of the screen.

3. Close the two boxes above the bottom informational box by dragging up the bar at the bottom center of each box.

4. Close Adobe Reader and ArcMap. You do not need to save anything.

Bonus exercises

As a bonus, additional exercises have been included to give you more GIS experience. These are not meant to be full chapters, but rather short exercises that explain some frequently sought after skill. It is assumed you have completed at least some of the chapters in this book and have acquired some facility with ArcGIS 10.1 for Desktop. Therefore, steps are not written in long detail, but rather in shorthand.

Bonus 1 Understanding common file types

Key concepts: understanding .shp, .mxd, .lyrs, .lpks, .mpk, basemap, and .gdb files

One of the most confusing things for new and old GIS hands alike is understanding ArcGIS 10.1 for Desktop file types. Here is a bonus exercise on how to make, save, and open common file types.

Files and tools

Files needed: All files for this exercise are accessible at C:\EsriPress\GIS20\B01.

Tools needed: ArcGIS 10.1 for Desktop and an Internet connection.

1 Open, copy, and save a shapefile (.shp)

A shapefile is one individual geographic layer that contains both the map and data. Shapefiles are a convenient and common file type, especially for basic mapping (see chapter 1 for more detail).

1. To open a shapefile, click the Add Data button, navigate to agejoined.shp, and click Add.

You cannot save a shapefile per se; however, you can make a copy of one and in so doing maintain changes to the attribute table and/or projection. The color of the shapefile is not saved during this process.

2. To make a copy of the file, right-click the layer name in the table of contents, click Data, and click Export Data. Navigate to your save folder. For Save as type, select Shapefile. Click OK.

2 Create, save, and open ArcMap documents (.mxd)

An ArcMap document (.mxd) is a pointer file to all the files used to create the map project. If you move any of the components to another place, ArcMap will ask you to reconnect to relevant files. Saving the ArcMap document does not save the shapefiles, but it does save the whole, overall project.

1. Select File > Save As. Navigate to your save folder, give the project a new name, and click Save.

2. Close ArcMap.

3. Open ArcMap. Select File > Open. Navigate to the .mxd, and click Open.

> ### INCREDIBLY USEFUL TIP
> *ArcMap documents can also be opened outside of ArcMap, in Windows Explorer, by double-clicking the document.*

3 Create, save, and open layer files (.lyr)

Layer files do not store data, but they can maintain the symbology and classification of the shapefile. You will need the shapefile that was used to create the layer file (or a copy of it) in order to open it.

1. Right-click the layer name in the table of contents and click Save as Layer File.

2. Select File > New > Blank Map > OK. You do not need to save any changes.

3. Click the Add Data button, navigate to the layer file, and open it.

> ### INCREDIBLY USEFUL TIP
> *If you have an open shapefile, from the Symbology options in its Properties, you can also import layer colors and symbols by clicking the Import button and selecting "Import symbology definition from another layer in the map or from a layer file."*

4 Create, save, and open layer packages (.lpk)

Layer packages (.lpk) save the layer and the shapefile. One advantage of this is that you can send just one layer with its associated coloring to someone instead of the whole project.

1. To create a layer package, right-click the layer name in the table of contents, click Create Layer Package, navigate to your save folder, and give the file a new name.

2. Close ArcMap. In Windows Explorer, navigate to the layer file and double-click it.

5 Create, save, and open map packages (.mpk)

A map package is a zipped version of the ArcMap document (.mxd). Map packages are useful for sharing your entire project with someone. To do this, perform the following steps:

1. Select File > Share As > Map Package. You can save to ArcGIS Online or to your desktop. Detailed steps for saving to ArcGIS Online are in chapter 20, and for saving to your desktop in chapter 19.

2. Close ArcMap. In Windows Explorer, navigate to the map package file and double-click it.

6 Open a basemap

Basemaps do not contain attribute information and are not shapefiles. They are visual images of maps; however, you cannot join to them or review an attribute table for them. Basemaps give contextual information to your map but are not very helpful for actual analysis.

1. To open a basemap, click the arrow beside the Add Data button. Click Add Basemaps and follow the prompts.

   NOTE: Issues have been reported while using browsers other than Internet Explorer.

7 Create and open a geodatabase (.gdb)

Geodatabases (.gdb) are like containers where you can store all files related to your GIS project. Grouping files together in one place is very convenient because it provides one point of access for all needed files. Chapter 15 goes into detail about creating geodatabases.

1. Open ArcCatalog. Right-click the save folder and select New > File Geodatabase. Give it a new name.

2. Select File > New > Blank Map > OK. You do not need to save any changes.

3. To open a geodatabase, either navigate to it through ArcCatalog or use the Add Data button.

Bonus 2 Advanced labeling

Key concepts: *making labels editable, labeling based on population, exploring new ArcGIS 10.1 for Desktop Maplex features*

When you place labels, they are affixed to the geography. So, you don't have much control over their placement. However, you may want to move them around or change the color and size of an individual label. You can do this by turning those labels into annotation (which makes individual labels into text boxes that you can move around and change).

Files and tools

Files needed: All files for this exercise are in C:\EsriPress\GIS20\B02.

Tools needed: ArcGIS 10.1 for Desktop.

Making labels editable

1 Label features in a shapefile

1. Add agejoinedcopy.shp to ArcMap. You will make irreparable changes to this file, so that is why we are using a copy of the agejoined file.

2. Right-click the layer name in the table of contents and click Properties. Click the Labels tab. Select the "Label features in this layer" box, and then click OK.

2 Convert to annotation

1. Right-click the layer name in the table of contents, click Convert Labels to Annotation. In the Store Annotation panel, click "In the map." Click Convert.

3 Move labels

1. Using the default pointer ▶ , move any label. Now delete a label.

2. Right-click a label, click Properties and Change Symbol. Scroll all the way down the preset styles to see three callout options (a square banner, a rounded banner, or just the leader line). Select one of these options. It will be necessary to move the label so you can see the anchor point for the callout.

UNPLACED LABELS

If you have too many labels, you will be shown a list of any unplaced labels in a window titled Overflow Annotation. These labels are not placed on the map. You can right-click these labels and select "Add annotation" to have them placed on the map. When all the labels are placed, this box will not appear.

Labeling based on population

Let's label cities with a population over 150,000 with Arial 12-point type in dark gray.

1 Open Cities.mxd

1. Select File > New > Blank Map > OK. You do not need to save any changes.
2. Select File > Open. Navigate to C:\EsriPress\GIS20\B02\Cities.mxd and click Open.
3. Right-click USCities, click Open Attribute Table, and notice the P001001 column (far right). This column contains the population for each city. Close the attribute table.

2 Turn on the Label toolbar

1. Select Customize > Toolbars > Labeling.
2. Click the Label Manager button. →
3. In the Label Classes section on the left, select the USCities box.
4. Click the USCities file name, and type **Major Cities** in the Enter Class Name field. Click Add. Notice that it has been added on the left. Clear the box for "default" and select the box for Major Cities. Click once on Major Cities and click the SQL Query button.

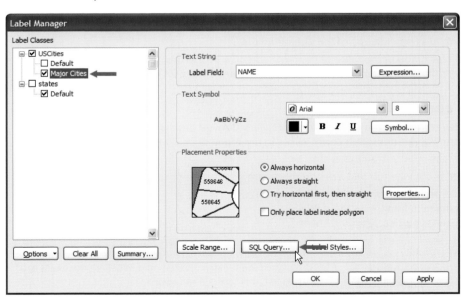

5. Write the query you see here. Click OK twice. Labels will be placed.
   "P001001" >=150000|

3 Give labels a halo

1. Click the Label Manager button 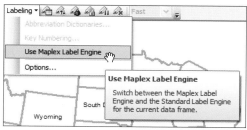 again, and then click the Major Cities name once.

2. Click Symbol > Edit Symbol > Mask > Halo > OK.

3. Change the font size to 12 points and bold the labels. Click OK twice.

New ArcGIS 10.1 for Desktop labeling features

4 Use Maplex Label Engine

Maplex Label Engine used to be sold as an extension of the core software; however, it is now included with all license levels of ArcGIS 10.1 for Desktop.

1. Right-click and remove the USCities layer.

2. Right-click states, click Properties, and click the Labels tab. Select the "Label features in this layer" box and click OK.

3. The Labeling toolbar should still be activated. If it is not, turn it on now.

4. Click states once in the table of contents to make it the active layer, and then click the Labeling button on the Labeling toolbar. Click Use Maplex Label Engine. ➜

5. Right-click the Symbol layer in the map table of contents. Click Properties, the Labels tab, and the Placement Properties button. Notice now many more options are available than under the regular menu.

6. On the Label Position tab, in the first field, select Land Parcel Placement. Click the Position button, click Curved, and click OK.

7. Switch to the Label Density tab and select the Remove Duplicates box. Also click Label Largest Feature Part. Click OK twice.

8. To go back and make adjustments, right-click the layer name in the table of contents and click Properties. Click the Labels tab and the Placement Properties button. There are many features to explore here and it is worth it to take a look around. You can also get a lot of information from the Help menu by searching for "Maplex."

9. Close ArcMap. You do not need to save anything.

Bonus 3 Creating custom symbology

Occasionally you may want to create your own symbols to symbolize features in maps. In this exercise, you will learn how to do that.

Files and tools

Files needed: All files for this exercise are in C:\EsriPress\GIS20\B03.

Tools needed: ArcGIS 10.1 for Desktop.

1 Create a new symbol

1. Add Symbols.shp to ArcMap.

2. Select Customize > Style Manager. Click the Styles button, and then click the Create New Style button. Call the new style **custom**. Click Save and OK. The new style should appear in the left panel of the Style Manager window.

3. Click the new style name once. Many options will appear on the right. The new .gif symbol type is marker. Right-click Marker Symbols, click New, and click Marker Symbols.

4. From the Type list, select Picture Marker Symbol. Navigate to C:\EsriPress\GIS20\ B03\custom.gif and click Open. Click OK (here you can give the new symbol a name) and Close.

> ### INCREDIBLY USEFUL TIP
> All standard picture types are supported, including JPEG.

5. Right-click the layer name in the table of contents, click Properties, and click the Symbology tab. Click the symbol swatch. The first symbol should be the new custom symbol. Select the symbol and click OK twice.

Bonus 4 Exporting to KML

KML stands for Keyhole Markup Language (Keyhole was the name of the application before Google bought it and added their own features and larger databases). KMZ stands for KML-Zipped. It is the default format for KML, and it is a compressed version of the file.

KMLs are used in many applications and are very popular for file sharing and working with multiple mapping applications.

Files and tools

Files needed: All files for this exercise are in C:\EsriPress\GIS20\B04.

Tools needed: ArcGIS 10.1 for Desktop.

GOOGLE MAPS SIZE LIMITATION

If you plan to upload a KMZ file to Google Maps, know that Google Maps has a file size limitation of 3 MB. Google Maps has a few other restrictions as well.

1 Open the project

1. Open ArcMap. Go to File > Open > Top25.mxd. This project includes 2010 population distribution by state and the top 25 "best places to live" in 2011 according to *Money* magazine.

LAYER TO KML

The Layer to KML tool exports an individual layer instead of all of the layers in the ArcMap document. Most of the options in this tool are the same as those in Map to KML.

2 Export map to KML

1. Open ArcToolbox 🖻 . Expand the Conversion Tools toolbox, and then expand the To KML toolset.

2. Double-click the Map to KML tool.

3. For the Map Document field, navigate to EsriPress\04\Top25.mxd. The Data Frame field will automatically fill in "Layers." For the Output File field, navigate to your save folder. Name the file **Top25.kmz**.

> **NOTE:** Do not type .kmz at the end of the file name. An error will be generated.

4. For the Map Output Scale field, type in the map's scale. The scale must be typed in with no commas and the 1: is not included. For example, a scale of 1:20,388,889 should be entered as 20388889. Leave everything else as it is and click OK.

EMBEDDING MAPS IN WEBSITES USING HTML

If you are able to upload a KMZ file to a web server, you can then access it using an Internet browser. Many popular online mapping applications give you the option of embedding the Iframe code into a website.

To make it work, copy the Iframe code into a blank HTML page, upload it to a server, and publish on your website.

> Paste HTML to embed in website
>
> `<iframe width="425" height="350" frameborder="0" :`
>
> Customize and preview embedded map

Bonus 5 Exploring Esri Maps for Office

Files and tools

Files needed: All files for this exercise are in C:\EsriPress\GIS20\B05.

Tools needed: ArcGIS 10.1 for Desktop, ArcGIS Online, and Microsoft Office 2010 or higher. Esri Maps for Office is a new ArcGIS feature and is an add-in to Microsoft Excel and PowerPoint 2010 or higher. It allows users to geocode from within these applications, and gives users some ability to create thematic maps.

> **NOTE:** Esri Maps for Office is a feature of ArcGIS Online. You will need an ArcGIS Online 30-day trial or paid subscription account to use Esri Maps for Office.

1 Set up a 30-day trial for ArcGIS Online (skip this step if you already have a subscription for your organization)

1. Navigate to http://www.esri.com. In the search field (upper, right) type **Esri Maps for Office**. Click the first link.
2. Set up a free 30-day trial for ArcGIS Online.
3. You will receive an e-mail with a download link for Esri Maps for Office.

2 Install Esri Maps for Office

1. Download the add-in software. Double-click the .exe file to begin installation.
2. Install any necessary prerequisites.

3 Open an Excel file

1. Open Excel 2010 (or higher).
2. Open popbystate.xlsx.
3. Put the cursor in cell A1.

4 Insert a map

1. Click the Esri Maps tab. Here you may be prompted for your ArcGIS Online user name and password. If so, enter this information. If not, click the Sign in button and enter your information.
2. On the Esri Map tab, in the Map group, click Insert Map. You must be in cell A1. →
3. Drag a corner of the map box to make the image larger.
4. Click the Esri Maps tab again.

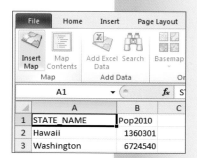

5 Add Excel data

1. On the Esri Maps tab, in the Add Data group, click Add Excel Data.

2. Click Cell Range as the option.

3. From cell A1, hold down the Shift key and use the right and down arrow keys on the keyboard to highlight the cell range. The cell range is $A$1:$B$52.

4. Click OK.

5. Select US State and click Next. →

6. Click the drop-down menu and select State_Name.

7. Click Add Data to Map. Scroll back up to the top of the spreadsheet, where a map displaying states will be visible.

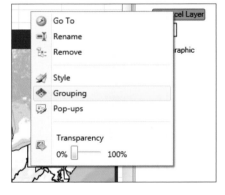

6 Thematically map

1. Right-click Excel Layer on the right side of the Excel window and select Grouping. →

2. Select Yes for "Do you want to group your data?"

3. For "Choose the column to group," select Pop2010 if it is not already selected.

4. Select a color from the color ramp. Leave all other options as they are. Click OK.

5. Reposition the map by clicking and moving it with your cursor. Use your mouse's scroll wheel to zoom in and out (or use the zoom tools in the upper left corner).

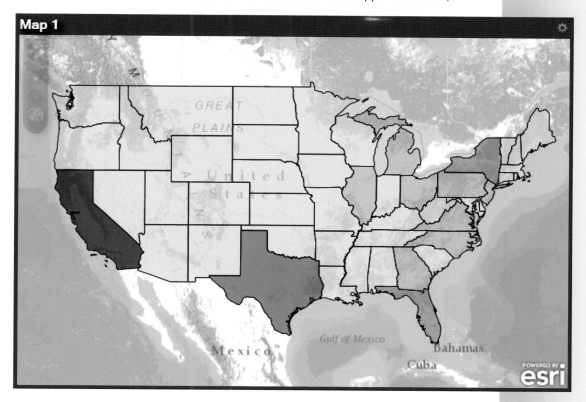

7 Give the map a title

1. Click the Settings button in the upper right corner of the map.

2. Type a title such as **Population by State, 2010**.

3. Click the Settings button again to go back to the map.

SHARING THE MAP

You can share the layer and map on ArcGIS Online by selecting those options on the Esri Maps tab. You can also use the Copy Image to Clipboard tool to copy and paste in another application like Microsoft Word. You can also use the Create Slide tool, which copies the image to PowerPoint.

8 Enable pop-up windows

1. Right-click Excel Layer on the right side of the Excel window and click Pop-ups.

2. Click OK.

3. Click any state and notice that a window pops up.

4. Close the window by clicking x in the upper right corner.

5. Close Excel. You do not need to save changes.

9 Geocode in Excel

1. Open Top25.xlsx.
2. Click the Esri Maps tab. Here you may be prompted for your ArcGIS Online user name and password. If so, enter this information.
3. On the Esri Maps tab, in the Map group, click Insert Map. You must be in cell A1.
4. Drag a corner of the map box to make the image larger.
5. Click the Esri Maps tab again.
6. In the Add Data group, click Add Excel Data.
7. Click Cell Range as the option.
8. From cell A1, hold the down Shift key and use the right and down arrow keys on the keyboard to highlight the cell range. The cell range is $A$1:$C$26.
9. Click OK.
10. Select US City and click Next.
11. Leave the default values.
12. Click Add Data to Map. Notice the little pushpins on the map. Scroll in to get a better view.

10 Fix errors

1. Notice on the right, under Excel Layers, it says 23 successful, 2 errors, Fix errors. Click Fix errors.
2. In the spreadsheet, type **New Brunswick** instead of South Brunswick. Click Enter and notice that one is geocoded.

GEOCODING ADDRESSES WITHIN EXCEL

In this example, we geocoded cities, but you could just as easily geocode addresses based on street address. Imagine quickly geocoding hundreds of clients to include a map with your mailing list. This is a very impressive addition to Excel.

The geocoder does not pick up Hunter's Creek, FL. An Internet search reveals that Hunter's Creek is near Kissimmee.

3. Type **Kissimmee** over Hunter's Creek and click Enter. It geocodes.

11 Change symbol

1. Right-click Excel Layers on the right.
2. Choose another symbol. Also, click Shapes at the top and notice the options there. Click OK.

12 Add other layers

You are able to add layers from ArcGIS Online or layers your organization maintains.

1. Click the Search tab on right side.
2. Type **Median Income by State** in the search field and search.
3. Down toward the bottom, there is a layer called USA Median Household Income 2011. Click the Add link for that layer.

4. Notice the more you zoom in, the smaller the geography gets. Zoom in until you can see data at the county level.

13 Reorder layers and close

1. In the Map Contents window on the right side, click Excel Layer (which is our geocoded layer) to activate it.

2. Click the Esri Maps tab again. In the Organize Layers group, click Bring Forward. This moves this layer into first position, making it visible on the map.

Here you have the same sharing options as above. Feel free to experiment with these.

3. Close Excel. You do not need to save.

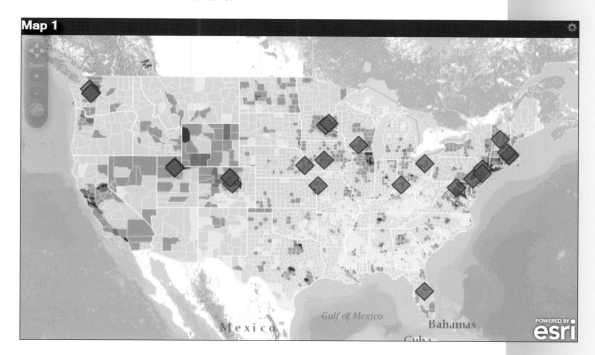

Bonus 6: Exploring metadata

Metadata provides your audience with key information about files and projects.

Files and tools

Files needed: All files are accessible in C:\EsriPress\GIS20\B06.

Tools needed: ArcGIS 10.1 for Desktop.

1 Choose a metadata style

1. The metadata style identifies the metadata standard you wish to follow. Click Customize > ArcMap Options > Metadata tab. Select FGDC CSDGM Metadata and click OK.

2 Update metadata

1. Open ArcCatalog and navigate to the place shapefile (tl_2011_01_place.shp) used in chapter 1 and also in C:\EsriPress\GIS20\B06.

2. Right-click the file name in ArcCatalog and click Item Description.

3. Toggle between the Description and Preview tabs at the top. One gives a text description of the file (in FGDC CSDGM metadata format) and the other visually displays the file.

4. Click the Edit button and notice all categories become editable. You do not need to edit anything here. Close.

METADATA STYLES

ArcMap provides a default style called Item Description. But, ArcGIS 10.1 supports many widely used metadata styles, including ISO 19139 Metadata Implementation Specification, North American Profile of ISO 19115 2003, INSPIRE Metadata Directive, and FGDC CSDGM Metadata. All metadata styles are very similar. For more information on this topic, type **Metadata Styles and Standards** into the Help menu search field.

IMPORTING METADATA STYLES

Using the Import button, you can either select a file that has a metadata style you would like to copy or a stand-alone XML file that contains the style.

Bonus 7: Working with multiple data frames

You can insert multiple data frames into one layout. This is done for many different purposes and is very useful to know.

Files and tools

Files needed: All files are accessible in C:\EsriPress\GIS20\B07.

Tools needed: ArcGIS 10.1 for Desktop.

1 Insert data frame

1. Open ArcMap. Go to File > Open > Top25.mxd.
2. Go to View > Layout view.
3. Go to Insert > Data frame. Notice a new empty frame has been added to your map. Drag the frame into an empty spot on your map so you can see it more clearly.

2 Add a shapefile to the new data frame

1. You can copy shapefiles from your original data frame into your new data frame. In the table of contents, right-click the Population shapefile and click Copy. Then, in the table of contents, right-click New Data Frame and click Paste Layer(s).

> **INCREDIBLY USEFUL TIP**
>
> Here you could have also clicked the data frame to make it active and used the Add Data button to insert shapefiles into the new data frame.

2. Now show only Alaska in the new frame. Click the new frame to make it the active layer. Click the Zoom In tool and draw a square around Alaska.
3. The new frame's background color is set to hollow. Give it a solid white fill. Click the frame to activate it. Little blue nodes will appear. Right-click the frame and click Properties > Frame tab. For background color, select a solid white fill. Click OK.

3 Add Extent Rectangle

You may want to do a "picture in picture" map, where one small frame relates to the larger map. To do this, perform the following steps:

1. Go to Insert > Data frame to add another data frame. Drag the frame into an empty spot on your map so you can see it more clearly.

2. In the table of contents, right-click the Population shapefile again and click Copy. Then, in the table of contents, right-click New Data Frame 2 and click Paste Layer(s).

3. Click the main map once to activate it and zoom in closely on Texas.

4. Right-click Data Frame 2 to activate it and click Properties.

5. Select the Extent Indicators tab and click Layers once under Other data frames. Use the right arrow to deposit it into the box on the right.

6. Select the Show Leader box and click OK. Notice Texas has a red box around it in Data Frame 2 showing the relationship between what's in the main map, and what's displayed in the smaller map.

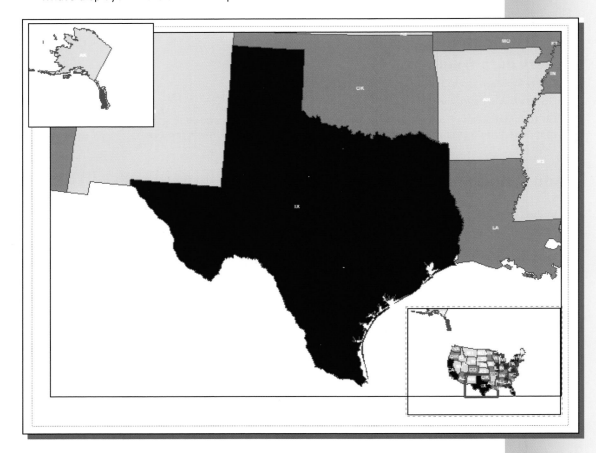

7. Close ArcMap.

Bonus 8: Enabling time

Time animation enables visualization of temporal analysis, or change over time. Using time animation tools in ArcMap, you are able to create thematic and geocoded maps that cycle through your information based on preset time intervals.

Files and tools

Files needed: All files are accessible in C:\EsriPress\GIS20\B08.

Tools needed: ArcGIS 10.1 for Desktop.

1 Open the project

1. Open ArcMap. Go to File > Open > HistPop.mxd. This project includes state populations from 1960 to 2010.

2 Open the attribute table

1. Right-click the layer name and click Open Attribute Table.

In order for the Time feature to work properly, every line you would like to be visible on the map must have a "time stamp" associated with it in the attributes table. Notice that each state is repeated six times, with the population for each decennial year from 1960 to 2010. In order to do a time animation with a thematic map, this is how the data must be laid out in the attribute tables (see illustration on the following page).

FID	Shape *	pop_OBJECT	pop_State	pop_Popula	pop_Date_
210	Polygon	1	Alabama	3,267,000	1960
211	Polygon	2	Alabama	3,444,000	1970
212	Polygon	3	Alabama	3,894,000	1980
213	Polygon	4	Alabama	4,040,000	1990
214	Polygon	5	Alabama	4,447,000	2000
215	Polygon	6	Alabama	4,780,000	2010
150	Polygon	7	Alaska	226,000	1960
151	Polygon	8	Alaska	303,000	1970
152	Polygon	9	Alaska	402,000	1980
153	Polygon	10	Alaska	550,000	1990
154	Polygon	11	Alaska	627,000	2000
155	Polygon	12	Alaska	710,000	2010
30	Polygon	13	Arizona	1,302,000	1960
31	Polygon	14	Arizona	1,775,000	1970
32	Polygon	15	Arizona	2,717,000	1980
33	Polygon	16	Arizona	3,665,000	1990
34	Polygon	17	Arizona	5,131,000	2000
35	Polygon	18	Arizona	6,392,000	2010
282	Polygon	19	Arkansas	1,786,000	1960
283	Polygon	20	Arkansas	1,923,000	1970
284	Polygon	21	Arkansas	2,286,000	1980
285	Polygon	22	Arkansas	2,351,000	1990
286	Polygon	23	Arkansas	2,673,000	2000
287	Polygon	24	Arkansas	2,916,000	2010
24	Polygon	25	California	15,717,000	1960
25	Polygon	26	California	19,971,000	1970
26	Polygon	27	California	23,668,000	1980
27	Polygon	28	California	29,758,000	1990
28	Polygon	29	California	33,872,000	2000
29	Polygon	30	California	37,254,000	2010
0	Polygon	31	Colorado	1,754,000	1960
1	Polygon	32	Colorado	2,210,000	1970
2	Polygon	33	Colorado	2,890,000	1980
3	Polygon	34	Colorado	3,294,000	1990
4	Polygon	35	Colorado	4,301,000	2000

2. Close the attribute table.

3 Enable time

1. Right-click the layer name in the table of contents and click Properties.
2. Click the Time tab and select the box to enable time on this layer.
3. Leave all settings as they are and click OK.

4 Play animation

1. Click the Time Slider button  to enable the Time Slider bar.

2. Click the Play button ▶ to view the animation. Notice how the map changes as it cycles through decades.

3. Click the Options button and notice the playback options and settings here. Click OK.

4. Click the Export to Video button and notice that you can export the animation as an AVI file, which could then be shared with others. Click OK.

5. Close ArcMap.

Bonus 9: Creating map books

The ability to create map books using functionality called Data Driven Pages was introduced in ArcGIS Desktop 10. In the next few steps, you will create a map book that consists of one PDF page for each county in a state.

Files and tools

Files needed: All files are accessible in C:\EsriPress\GIS20\B09.

Tools needed: ArcGIS 10.1 for Desktop.

1 Open the project

1. Open ArcMap. Go to File > Open > MapBook.mxd.

2 Switch to layout view

1. On the View menu, click Layout View. It is best to create a map book from the layout view.

2. On the Layout toolbar, click the Zoom to Whole Page button ⊡ on the Layout toolbar so you can see the full layout. If your map isn't centered, feel free to reposition it using the Pan tool, but it is not necessary.

3 Create a map book using Data Driven Pages

1. Using the default pointer, select the title and press the Delete key on your keyboard to delete the title. Select the source text and delete that as well. Only the map and legend should be included in the layout.

2. Go to Customize > Toolbars > Data Driven Pages. A new toolbar will be added to the window.

3. Click the Page Text button, and then click Data Driven Page Name.

4. A very small text box will appear in the center of the map. Drag the text box to the upper left corner of the map, right-click it, and click Properties. This will allow you to have a title on each page with the county name.

In the attribute table there are two columns that would work well for page titles: NAME or NAMELSAD. NAME has just the county name without the word "County" behind it. The column NAMELSAD, however, contains both the county name and the word county behind it. Let's use the NAMELSAD column to create the titles.

5. Right-click the title text box and replace Name with **NAMELSAD**. →

6. Click Change Symbol and make the font bold and size 22. Click OK twice.

```
Text:
<dyn type="page" property="namelsad"/>
```

7. Click the Data Driven Page Setup button 🖼. It might be difficult to see, but it is the first button on the new toolbar.

8. Select the Enable Data Driven Pages box.

9. For the Layer Field, select AgeJoined.

10. For the Name Field, select Name. This is the column that determines what field will be used to create the various maps. We have selected Name, which is the name of each county, so each county gets a page. Also, change the Sort field to County, so the pages will be alphabetically sorted by county name. Leave all other default settings as they are. Click OK.

11. Review the results. Scroll through the pages of the new map book by clicking the blue arrows to scroll to the next page.

4 Create a PDF of the map book

1. On the File menu, click Export Map.

2. If the Save as type is not already PDF, select PDF from the list.

3. Click the Pages tab.

4. This map book should have 67 pages (one for each county). Let's print only pages one through five. Select the Page Range option and type **1–5**.

5. Click Save.

6. Open MapBook.pdf and review the results. These are geoenabled PDFs (covered in chapter 20).

SOURCE CREDITS

Data credits

Chapter 1

\EsriPress\GIS20\01\alabamacounties.shp, courtesy of US Census Bureau.

\EsriPress\GIS20\01\mystate.mxd, created by the author from US Census Bureau data.

\EsriPress\GIS20\01\tl_2011_01_place.shp, courtesy of US Census Bureau.

\EsriPress\GIS20\01\tl_2011_01_us_county.shp, courtesy of US Census Bureau.

Chapter 2

\EsriPress\GIS20\02\alabamacounties.shp, courtesy of US Census Bureau.

\EsriPress\GIS20\02\mystate.mxd, created by the author from US Census Bureau data.

\EsriPress\GIS20\02\refmap.mxd, created by the author from US Census Bureau data.

\EsriPress\GIS20\02\tl_2011_01_place.shp, courtesy of US Census Bureau.

Chapter 3

\EsriPress\GIS20\03\alabamacounties.shp, courtesy of US Census Bureau.

\EsriPress\GIS20\03\countiesprj.shp, created by the author from US Census Bureau data.

Chapter 4

\EsriPress\GIS20\04\age.xlsx, created by the author from US Census Bureau data.

\EsriPress\GIS20\04\DEC_10_DP_DPDP1_with_ann.csv, courtesy of US Census Bureau.

Chapter 5

\EsriPress\GIS20\05\age.xlsx, created by the author from US Census Bureau data.

\EsriPress\GIS20\05\agejoined.shp, created by the author from US Census Bureau data.

\EsriPress\GIS20\05\countiesprj.shp, created by the author from US Census Bureau data.

Chapter 6

\EsriPress\GIS20\06\agejoined.shp, created by the author from US Census Bureau data.

\EsriPress\GIS20\06\seniors.mxd, created by the author from US Census Bureau data.

Chapter 7

\EsriPress\GIS20\07\agejoined.shp, created by the author from US Census Bureau data.

\EsriPress\GIS20\07\minifile.shp, created by the author from US Census Bureau data.

Chapter 8

\EsriPress\GIS20\08\agencies.mxd, created by the author from US Census Bureau data.

\EsriPress\GIS20\08\Agencies.xls, USADATA courtesy of Dun and Bradstreet.

\EsriPress\GIS20\08\Geocoding_Result, created by the author from USADATA courtesy of Dun and Bradstreet.

\EsriPress\GIS20\08\tl_2011_48209_areawater.shp, courtesy of US Census Bureau.

EsriPress\GIS20\08\tl_2011_48209_edges.shp, courtesy of US Census Bureau.

Chapter 9

\EsriPress\GIS20\09\agencies.mxd, created by the author from US Census Bureau data.

\EsriPress\GIS20\09\tl_2011_48029_areawater.shp, courtesy of US Census Bureau.

\EsriPress\GIS20\09\tl_2011_48029_edges.shp, courtesy of US Census Bureau.

Chapter 10

\EsriPress\GIS20\10\LatLongsPlotted.shp, USADATA courtesy of Dun and Bradstreet.

\EsriPress\GIS20\10\tl_2011_48029_edges.shp, courtesy of US Census Bureau.

\EsriPress\GIS20\10\XYData.xls, USADATA courtesy of Dun and Bradstreet.

Chapter 11

\EsriPress\GIS20\11\newsites.shp, created by the author from US Census Bureau data.

\EsriPress\GIS20\11\southwest.shp, created by the author from US Census Bureau data.

\EsriPress\GIS20\11\states.shp, courtesy of US Census Bureau.

Chapter 12

\EsriPress\GIS20\12\agejoined.shp, created by the author from US Census Bureau data.

\EsriPress\GIS20\12\seniors.shp, created by the author from US Census Bureau data.

Chapter 13

\EsriPress\GIS20\13\seniors.shp, created by the author from US Census Bureau data.

\EsriPress\GIS20\13\tl_2011_01_place.shp, courtesy of US Census Bureau.

Chapter 14

\EsriPress\GIS20\14\buffers.shp, created by the author from US Census Bureau data.

\EsriPress\GIS20\14\circle.shp, created by the author from US Census Bureau data.

\EsriPress\GIS20\14\divisions.shp, created by the author from US Census Bureau data.

\EsriPress\GIS20\14\PNW.shp, created by the author from US Census Bureau data.

\EsriPress\GIS20\14\states.shp, courtesy of US Census Bureau.

\EsriPress\GIS20\14\SW.shp, created by the author from US Census Bureau data.

\EsriPress\GIS20\14\tl_2011_48029_edges.shp, courtesy of US Census Bureau.

\EsriPress\GIS20\14\youthcenters.shp, USADATA courtesy of Dun and Bradstreet.

Chapter 15

\EsriPress\GIS20\15\gdb.gdb, created by the author from US Census Bureau data.

\EsriPress\GIS20\15\agejoined.shp, created by the author from US Census Bureau data.

\EsriPress\GIS20\15\Agencies.xls, USADATA courtesy of Dun and Bradstreet.

\EsriPress\GIS20\15\Alamo.jpg, courtesy of DigitalGlobe.

\EsriPress\GIS20\15\tl_2011_48029_edges.shp, courtesy of US Census Bureau.

Chapter 16

\EsriPress\GIS20\16\alabamacounties.shp, created by the author from US Census Bureau data.

\EsriPress\GIS20\16\spatialjoin.shp, created by the author from US Census Bureau data.

\EsriPress\GIS20\16\tl_2011_01_place.shp, courtesy of US Census Bureau.

Chapter 17

\EsriPress\GIS20\17\Alamo.jpg, courtesy of DigitalGlobe.

\EsriPress\GIS20\17\tl_2011_48029_edges.shp, courtesy of US Census Bureau.

Chapter 18

\EsriPress\GIS20\18\agejoined.shp, created by the author from US Census Bureau data.

Chapter 19

\EsriPress\GIS20\19\agejoined.lpk, created by the author from US Census Bureau data.

\EsriPress\GIS20\19\agejoined.shp, created by the author from US Census Bureau data.

\EsriPress\GIS20\19\circle.shp, created by the author from US Census Bureau data.

\EsriPress\GIS20\19\Clip.gpk, created by the author from US Census Bureau data.

\EsriPress\GIS20\19\clippedagain.shp, created by the author from US Census Bureau data.

\EsriPress\GIS20\19\seniors.mpk, created by the author from US Census Bureau data.

\EsriPress\GIS20\19\seniors.mxd, created by the author from US Census Bureau data.

\EsriPress\GIS20\19\states.shp, courtesy of US Census Bureau.

Chapter 20

\EsriPress\GIS20\20\AgeJoined.shp, created by the author from US Census Bureau data.

\EsriPress\GIS20\20\seniors.mpk, created by the author from US Census Bureau data.

\EsriPress\GIS20\20\seniors.mxd, created by the author from US Census Bureau data.

Bonus Exercises

\EsriPress\GIS20\B01\agejoined.shp, created by the author from US Census Bureau data.

\EsriPress\GIS20\B02\agejoinedcopy.shp, created by the author from US Census Bureau data.

\EsriPress\GIS20\B02\cities.mxd, created by the author from US Census Bureau data.

\EsriPress\GIS20\B02\states.shp, courtesy of US Census Bureau.

\EsriPress\GIS20\B02\USCities.shp, courtesy of US Census Bureau.

\EsriPress\GIS20\B03\custom.gif, created by the author.

\EsriPress\GIS20\B03\Symbols.shp, created by the author.

\EsriPress\GIS20\B04\Population.shp, created by the author from US Census Bureau data.

\EsriPress\GIS20\B04\Top25.mxd, created by the author from US Census Bureau data.

\EsriPress\GIS20\B04\Top25.shp, created by the author from US Census Bureau data.

\EsriPress\GIS20\B05\Completedpopbystate.xlsx, created by the author from US Census Bureau data.

\EsriPress\GIS20\B05\CompletedTop25.xlsx, created by the author from US Census Bureau data.

\EsriPress\GIS20\B05\popbystate.xlsx, created by the author from US Census Bureau data.

\EsriPress\GIS20\B05\Top25.xlsx, created by the author from US Census Bureau data.

\EsriPress\GIS20\B06\tl_2011_01_place.shp, courtesy of US Census Bureau.

\EsriPress\GIS20\B07\Population.shp, created by the author from US Census Bureau data.

\EsriPress\GIS20\B07\Top25.mxd, created by the author from US Census Bureau data.

\EsriPress\GIS20\B07\Top25.shp, created by the author from US Census Bureau data.

\EsriPress\GIS20\B08\HistoricalPop.shp, created by the author from US Census Bureau data.

\EsriPress\GIS20\B08\HistPop.mxd, created by the author from US Census Bureau data.

\EsriPress\GIS20\B09\AgeJoined.shp, created by the author from US Census Bureau data.

\EsriPress\GIS20\B09\MapBook.mxd, created by the author from US Census Bureau data.

Figure Credits

Chapter 3

Chapter 3, graphic of UTM zones courtesy of the USGS.

Chapter 3, graphic of state plane coordinate system created by the author from files in ArcGIS.

DATA LICENSE AGREEMENT

Important: Read carefully before opening the sealed media package.

Environmental Systems Research Institute, Inc. (Esri) is willing to license the enclosed data and related materials to you only upon the condition that you accept all of the terms and conditions contained in this license agreement. Please read the terms and conditions carefully before opening the sealed media package. By opening the sealed media package, you are indicating your acceptance of the Esri License Agreement. If you do not agree to the terms and conditions as stated, then Esri is unwilling to license the data and related materials to you. In such event, you should return the media package with the seal unbroken and all other components to Esri.

Esri License Agreement

This is a license agreement, and not an agreement for sale, between you (Licensee) and Environmental Systems Research Institute, Inc. (Esri). This Esri License Agreement (Agreement) gives Licensee certain limited rights to use the data and related materials (Data and Related Materials). All rights not specifically granted in this Agreement are reserved to Esri and its Licensors.

Reservation of Ownership and Grant of License: Esri and its Licensors retain exclusive rights, title, and ownership to the copy of the Data and Related Materials licensed under this Agreement and, hereby, grant to Licensee a personal, nonexclusive, nontransferable, royalty free, worldwide license to use the Data and Related Materials based on the terms and conditions of this Agreement. Licensee agrees to use reasonable effort to protect the Data and Related Materials from unauthorized use, reproduction, distribution, or publication.

Proprietary Rights and Copyright: Licensee acknowledges that the Data and Related Materials are proprietary and confidential property of Esri and its Licensors and are protected by United States copyright laws and applicable international copyright treaties and/or conventions.

Permitted Uses: Licensee may install the Data and Related Materials onto permanent storage device(s) for Licensee's own internal use. Licensee may make only one (1) copy of the original Data and Related Materials for archival purposes during the term of this Agreement unless the right to make additional copies is granted to Licensee in writing by Esri. Licensee may internally use the Data and Related Materials provided by Esri for the stated purpose of GIS training and education.

Uses Not Permitted: Licensee shall not sell, rent, lease, sublicense, lend, assign, time-share, or transfer, in whole or in part, or provide unlicensed Third Parties access to the Data and Related Materials or portions of the Data and Related Materials, any updates, or Licensee's rights under this Agreement. Licensee shall not remove or obscure any copyright or trademark notices of Esri or its Licensors.

Term and Termination: The license granted to Licensee by this Agreement shall commence upon the acceptance of this Agreement and shall continue until such time that Licensee elects in writing to discontinue use of the Data or Related Materials and terminates this Agreement. The Agreement shall automatically terminate without notice if Licensee fails to comply with any provision of this Agreement. Licensee shall then return to Esri the Data and Related Materials. The parties hereby agree that all provisions that operate to protect the rights of Esri and its Licensors shall remain in force should breach occur.

Disclaimer of Warranty: The Data and Related Materials contained herein are provided "as is," without warranty of any kind, either express or implied, including, but not limited to, the implied warranties of merchantability, fitness for a particular purpose, or noninfringement. Esri does not warrant that the Data and Related Materials will meet Licensee's needs or expectations, that the use of the Data and Related Materials will be uninterrupted, or that all nonconformities, defects, or errors can or will be corrected. Esri is not inviting reliance on the Data or Related Materials for commercial planning or analysis purposes, and Licensee should always check actual data.

Data Disclaimer: The Data used herein has been derived from actual spatial or tabular information. In some cases, Esri has manipulated and applied certain assumptions, analyses, and opinions to the Data solely for educational training purposes. Assumptions, analyses, opinions applied, and actual outcomes may vary. Again, Esri is not inviting reliance on this Data, and the Licensee should always verify actual Data and exercise their own professional judgment when interpreting any outcomes.

Limitation of Liability: Esri shall not be liable for direct, indirect, special, incidental, or consequential damages related to Licensee's use of the Data and Related Materials, even if Esri is advised of the possibility of such damage.

No Implied Waivers: No failure or delay by Esri or its Licensors in enforcing any right or remedy under this Agreement shall be construed as a waiver of any future or other exercise of such right or remedy by Esri or its Licensors.

Order for Precedence: Any conflict between the terms of this Agreement and any FAR, DFAR, purchase order, or other terms shall be resolved in favor of the terms expressed in this Agreement, subject to the government's minimum rights unless agreed otherwise.

Export Regulation: Licensee acknowledges that this Agreement and the performance thereof are subject to compliance with any and all applicable United States laws, regulations, or orders relating to the export of data thereto. Licensee agrees to comply with all laws, regulations, and orders of the United States in regard to any export of such technical data.

Severability: If any provision(s) of this Agreement shall be held to be invalid, illegal, or unenforceable by a court or other tribunal of competent jurisdiction, the validity, legality, and enforceability of the remaining provisions shall not in any way be affected or impaired thereby.

Governing Law: This Agreement, entered into in the County of San Bernardino, shall be construed and enforced in accordance with and be governed by the laws of the United States of America and the State of California without reference to conflict of laws principles. The parties hereby consent to the personal jurisdiction of the courts of this county and waive their rights to change venue.

Entire Agreement: The parties agree that this Agreement constitutes the sole and entire agreement of the parties as to the matter set forth herein and supersedes any previous agreements, understandings, and arrangements between the parties relating hereto.

INDEX